HARD COAL, HARD TIMES:
Ethnicity and Labor in the Anthracite Region

Edited by David L. Salay

The SCRANTON ANTHRACITE MUSEUM ASSOCIATES
in cooperation with the
Commonwealth of Pennsylvania
PENNSYLVANIA HISTORICAL AND MUSEUM
COMMISSION

The Anthracite Museum Press
Scranton, 1984

© 1983 by the Scranton Anthracite Museum Associates
R. D. 1 Bald Mountain Road
Scranton, Pennsylvania 18504

Library of Congress Cataloging in Publication Data

Salay, David L. 1944--
  Ethnicity and Labor in the Anthracite Region.

1. Pennsylvania—Foreign Population—Addresses, essays, lectures.
2. Technology—Pennsylvania—Addresses, essays, lectures.

I. Title.

ISBN: 0-917445-00-7

Printed in the United States of America

SUPPORTED BY A GRANT FROM:

A STATEWIDE FUNDING ORGANIZATION
FUNDED IN PART BY THE NATIONAL
ENDOWMENT FOR THE HUMANITIES

This volume was published by the Scranton Anthracite Museum Associates with the cooperation and support of the Pennsylvania Historical and Museum Commission in its continuing effort to preserve the history of the people of the Commonwealth. The symposium, which was the basis for this volume, and this publication were funded in part by a grant from the Pennsylvania Humanities Council, a statewide organization funded in part by the National Endowment for the Humanities. The Scranton Anthracite Museum Associates and the Pennsylvania Historical and Museum Commission provided matching grants for the publication.

Opinions and ideas expressed in this volume do not necessarily represent the views of the Pennsylvania Historical and Museum Commission, the Scranton Anthracite Museum Associates, the Pennsylvania Humanities Council, or any agency of the Federal Government.

THE PENNSYLVANIA HISTORICAL AND MUSEUM COMMISSION

Mrs. Frank N. Piasecki, Chairman

D. David Eisenhower
Mrs. J. Welles Henderson
Leroy Patrick
Mrs. Russell D. Robison

Mrs. Robert S. Ross
Maxwell Whiteman
Mrs. F. Karl Witherow

Arthur P. Ziegler, Jr.

MEMBERS FROM THE GENERAL ASSEMBLY

Richard A. Snyder, Senator
Hardy Williams, Senator
James L. Wright, Representative
Kurt D. Zwikl, Representative

MEMBERS, EX-OFFICIO

Robert C. Wilburn, Secretary of Education
Clyde M. McGeary, Designate

ADMINISTRATIVE STAFF

Larry E. Tise, Executive Director
Nancy Kolb, Assistant Executive Director
Robert Sieber, Acting Director, Bureau of Museums

SCRANTON ANTHRACITE MUSEUM ASSOCIATES

Mrs. Adrienne Horger, President   Mr. John Cognetti, Vice President
Mr. John Scheuer, Secretary   Mr. William Smith, Treasurer

BOARD OF DIRECTORS

Mr. William Dawson
Mr. John Hennemuth
Mr. Roger Howell
Mr. David Leung
Mrs. Linda Lynett
Mr. George Nichols
Mrs. Anne Powell
Mr. Tom Reddington
Mr. Gerald Scofield
Mrs. Stephanie Swatkowski
Mrs. Karen Welles

CONTENTS

Contributors	iv
Editor's Introduction	vi

1. The Miners of St. Clair:
   Family, Class, and Ethnicity in a Mining Town in Schuylkill County, 1850-1880
   Anthony F. C. Wallace 1
2. Themes from Immigrant Fraternal Life:
   The Early Decades of the Hazleton Based Hungarian Verhovay Sick Benefit Association
   Bela Vassady, Jr. 17
3. Commentary:
   The Miners of St. Clair and Themes from Immigrant Fraternal Life
   Joseph T. Makarewicz 35
4. Corporate Attitudes Toward Labor Organizations:
   The Controversy over the Price of Powder in the Lackawanna Valley, 1888-1889
   Perry K. Blatz 40
5. Hendrick B. Wright:
   "A Practical Treatise on Labor"
   James P. Rodechko 58
6. Commentary:
   Corporate Attitudes Toward Labor Organizations and Hendrick B. Wright
   Melvyn Dubofsky 75
7. The Family Economy and Labor Protest in Industrial America
   John Bodnar 78
8. The Coal and Iron Police in Anthracite Country
   Stephen R. Couch 100
9. Commentary:
   The Family Economy and Labor Protest in Industrial America and The Coal and Iron Police in Anthracite Country
   Ronald M. Benson 120
10. Ethnic Responses to the Lattimer Massacre
    George A. Turner 126

11. "Do Your Duty!":
    Editorial Response to the Anthracite Strike of 1902
    Harold W. Aurand   153
12. Commentary:
    Ethnic Responses to the Lattimer Massacre and "Do Your Duty"
    Ronald L. Filippelli   165
13. The Ethno-Religious Factor Reaches Fruition:
    The Politics of Hard Coal, 1945–1972
    William A. Gudelunas   169
14. Commentary:
    The Ethno-Religious Factor Reaches Fruition
    Matthew S. Magda   189

CONTRIBUTORS

Harold W. Aurand, Associate Professor of History, The Pennsylvania State University, Hazleton Campus, Hazleton, Pennsylvania

Ronald M. Benson, Associate Professor of History, Millersville State University, Millersville, Pennsylvania

Perry K. Blatz, Research Associate, New Jersey Historical Commission, Trenton, New Jersey

John Bodnar, Associate Professor of History, Indiana University, Bloomington, Indiana

Stephen R. Couch, Assistant Professor of Sociology, The Pennsylvania State University, Schuylkill Campus, Schuylkill Haven, Pennsylvania

Melvyn Dubofsky, Professor of History, State University of New York at Binghamton, Binghamton, New York

Ronald L. Filippelli, Professor and Head, Department of Labor Studies, The Pennsylvania State University, University Park, Pennsylvania

William A. Gudelunas, Associate Professor of History, The Pennsylvania State University, Schuylkill Campus, Schuylkill Haven, Pennsylvania

Matthew S. Magda, Associate Historian, Division of History, Pennsylvania Historical and Museum Commission, Harrisburg, Pennsylvania

Joseph T. Makarewicz, Director, Pennsylvania Ethnic Heritage Studies Center, University of Pittsburgh, Pittsburgh, Pennsylvania

James P. Rodechko, Professor of History, Wilkes College, Wilkes-Barre, Pennsylvania

George A. Turner, Associate Professor of History, Bloomsburg, State University, Bloomsburg, Pennsylvania

Bela Vassady, Jr., Associate Professor of History, Elizabethtown College, Elizabethtown, Pennsylvania

Anthony F. C. Wallace, Professor of Anthropology, University of Pennsylvania, Philadelphia, Pennsylvania

EDITOR'S INTRODUCTION

The Anthracite Region of Pennsylvania is a well defined geological and cultural area. Within 484 square miles of northeastern Pennsylvania is found almost the entire supply of anthracite*, or hard coal in the United States. The concentration of this mineral resource dominated the economy and culture of the area. The Anthracite Region became known both for the all encompassing presence of the coal mining industry and the strength of the region's ethnic communities.

The development of the Anthracite Region began in the eighteenth century with agricultural settlements and iron manufactories. But in the 19th century, the coal industry—anthracite mining—became the basis for the economic and cultural growth of northeastern Pennsylvania. Collieries and railroads dominated life in the Anthracite Region for over 100 years. Villages and cities were built to house and service the population which came to mine, process, and transport anthracite. Industries grew up to support the coal industry or to use the inexpensive fuel and the labor available from the miners' wives and children.

The coal companies encouraged immigration to the Anthracite Region and actively sought the manpower needed to serve their coal mining operations. Because of this, there was a direct link between labor and ethnic development in the Anthracite Region. In some respects the Region was a microcosm of immigrant settlement patterns in the United States. Yet, it also represented a distinctly regional development because of its dependence on the seasonal and economic fluctuations of a single industry. Mining conditions, mining accidents, and mining disasters affected the entire community. Miners often lived in company houses, shopped in company stores, attended company maintained schools, and were treated by company doctors. The colliery operations and their resulting refuse piles, surface mining pits, and mine subsidence shaped the environment in which the miners lived. The harsh working conditions, the isolation of many communities, and the coal companies' efforts to play one ethnic group against another resulted in strong family and community enclaves.

During the 1920's, the importance of anthracite began to decline with the

---

*Anthracite is a ranking of coal. A ranking is based on the carbon content and percentage of volatile material. The high percentage of carbon in anthracite, its low percentage of volatile material, and the low sulfur content, make it a slow burning, clean, high BTU value fuel.

switch from coal to oil and gas for home heating and from steam engines to diesel engines and electric motors. Changes in fuel consumption along with the expansion of mechanized surface mining operations and the reduction in underground mining resulted in miner layoffs. The decline in the economic base, along with changes in government assistance and consumer technology, altered ethnic and community patterns.

The decline in anthracite's importance as a fuel coupled with an unfavorable view of northeastern Pennsylvania as a "coal cracker" country resulted in the negative self-image of the Region, its history, and its cultural heritage. Memories of harsh economic times and of deaths or injuries resulting from work in the mines seemed to be part of a heritage worth forgetting. Nevertheless, a community and its culture is the result of a history that cannot be altered in fact.

At the same time, there are accomplishments and traditions worth celebrating and there are residents who have strong positive impressions of their life in the Anthracite Region and pride in their coal mining heritage. They note that later generations benefited from the labor and efforts of those who came before and took advantage of available opportunities. The continuation or revival of ethnic traditions and an interest in local history are the result of a growing awareness of the important role the Anthracite Region played in the lives of the people who lived here and in the development of Pennsylvania and the United States.

This awareness received timely support by the Legislature of the Commonwealth of Pennsylvania. In 1971, the Legislature created the Pennsylvania Historical and Museum Commission's Anthracite Museum Complex to preserve and interpret the history, technology, and culture of the Anthracite Region. In part it may be considered a response to the calls by S. K. Stevens, Philip Klein, and J. Cutler Andrews in the 1950's and 1960's for more studies of regional and local history, and for studies of ethnic groups.

The Anthracite Museum Complex was designed to encompass all facets of life in the region and serve as an educational institution for area residents and for visitors to northeastern Pennsylvania. Each of the four units focuses on a particular segment of the Region's history. The Ashland Anthracite Museum focuses on mining technology: the exploration, extraction, and processing of anthracite. Eckley Miners' Village, a mid-nineteenth century mining community, features the everyday life of the miner and his family. The Scranton Anthracite Museum provides an overview of the economic and institutional development of the Anthracite Region with themes that include immigration, transportation, the growth of villages and cities, and the formation of social and labor organizations. The remaining furnace stacks of the Lackawanna

Iron and Coal Company, once the second largest producer of iron in the United States, represent one of the important markets for anthracite. These museums and historic sites provide a comprehensive overview of the Anthracite Region and serve as repositories for documents and artifacts.

A series of public lectures, films, and workshops have been instituted to encourage interest in the history of the Anthracite Region. To encourage scholarly research of the Anthracite Region, a series of conferences and symposia were established. The first symposium, on "Ethnicity and Labor in the Anthracite Region," was the basis for this publication. During the symposium the formal presentation of papers was followed by a commentator's critical review, and then a general discussion between the audience and the speakers. Some of the comments and suggestions made during these discussions have been incorporated in the papers published here.

The papers presented at this conference make several points. The first is that there is a growing scholarly interest in the Anthracite Region. A second, as the title indicates, is the importance of understanding the interaction and inter-relationship between ethnicity and labor in the Anthracite Region. The third, as a review of the chapters in this volume indicate, are the various areas of research now being undertaken and the possibilities for further regional studies.

The issues studied by these authors touch on some of the central concerns of ethnic studies and labor history. The question of cultural pluralism and ethnic conflicts are raised in Anthony F. C. Wallace's chapter on "The Miners of St. Clair." In a review of several key areas—family, occupation, and ethnicity—Wallace points out that religious and ethnic differences were less important than the common cultural elements found within the region. He concludes that, in labor terms, the image of violent ethnic and religious rivalries was more often the product of mine operator propaganda than of actual life in the Anthracite Region.

William A. Gudelunas' "The Ethno-Religious Factor Reaches Fruition," on the other hand, sees ethnic and religious factors as features in the political battles of Schuylkill County through 1972. Gudelunas explains this persistence of ethno-religious voting patterns in terms of the county's position as an economic "internal colony" with absentee owners controlling the industrial base and a declining population which did not bring new voters into the county. Gudelunas sees the collapse of the anthracite industry as a factor in the continuation of the political rivalry expressed by the traditional German-Protestant versus Irish Catholic alignments.

Conflict could exist not only between ethnic groups but within them as well. Because nineteenth century immigrants could not depend on government assistance or company welfare programs if illness struck, they formed benefit societies to protect themselves and their families. Many of these societies were short-lived, but some, like the Hungarian Verhovay Sick Benefit Association, developed into professional life insurance companies. Bela Vassady, Jr.'s "Themes from Immigrant Fraternal Life" analyzes the conflict generated with-in the Association between those who viewed it in traditional, Magyar terms and those who wanted to broaden the ethnic base of the Association and apply modern business techniques to its operation.

Although nineteenth century America propounded the philosophy of uplift and social improvement, the industrialists and mine owners were more interested in the immigrants for the labor they could perform. Perry K. Blatz's "Corporate Attitudes Toward Labor Organizations" shows that the mine owners' paternalistic concern was related to their control of the working place and was often a front to retain power over wages, hours, and working conditions. Blatz's analysis of the controversy over the price miners paid for their blasting powder illustrates how mine owners could use their economic control to stifle unionization efforts. Blatz concludes that the mine owners' actions convinced the miners that paternalism was only a mask to keep them from organizing.

In addition to economic controls, the mine owners also used more direct means of coercion. Stephen R. Couch's "The Coal and Iron Police in Anthracite Country" deals with a unique police force licensed by the Commonwealth of Pennsylvania between 1866 and 1935. Although paid and controlled by the coal companies, the Coal and Iron Police were the primary law enforcement agencies in parts of the Anthracite Region. Their presence helped the coal companies to exercise social control of the area. Couch argues that the Coal and Iron Police were developed because of the conditions related to rural industrialization at that time. He concludes that while they were used successfully by the coal companies at the outset, the Coal and Iron Police indirectly helped encourage the development of a strong labor organization in the Anthracite Region.

The struggle for union organization has been marred at times by violence and murder. In the Anthracite Region this is most often associated with the miners in the guise of the "Molly Maguires." One of the greatest tragedies, however, was executed by the mine owners. The Lattimer Massacre of 1897 resulted in the wounding of thirty-eight Slavic miners and the death of nineteen. George A. Turner's "Ethnic Responses to the Lattimer Massacre" reviews the public reaction to what was one of the worst tragedies in American labor history. Turner finds that unlike mine owners had predicted, the workers' reaction to the massacre was not more violence but rather a concerted effort for legal redress and compensation. Yet, even after the sheriff and his deputies were exonerated and compensation was denied, new violence did not erupt.

The miners' attempts to unionize did receive occasional support from community leaders in the nineteenth century. James P. Rodechko's "Hendrick B. Wright's A PRACTICAL TREATISE ON LABOR" analyzes the reasoning behind this support. Rodechko points out that while Wright was accused of trying to further his own political fortunes, he really was writing in support of traditional middle class values, values that applied equally to the working class. Wright's defense of the miners' need to organize themselves for protection was in part an attack on the encroachments of national monopolists and a defense of local entrepreneurs. Rodechko concludes that Wright hoped to use political action and worker cooperation to both restrain these outsiders and uplift the laboring classes by returning to what he viewed as the traditional American values.

Harold W. Aurand's "Do Your Duty" reviews the support received by the 1902 mine strikers from a different group. While news reporters covered the daily events, newspaper editors provided a broader analysis. Initially, editorial comment warned of the possibility of a strike and the dangers it represented. After the strike began both labor and management were criticized. Editors condemned the mine owners more, however. This was not for their use of strikebreakers or for exploiting the workers but rather for the owners' failure to manage what they viewed as a public trust. When the mine owners refused to negotiate with the unions, the editors demanded federal intervention. Aurand concludes that the editors viewed the mine operators as having the primary responsibility for providing coal to the public. They held the mine owners accountable for failing to meet the public's need.

While labor protest continued to use the strike as an organizing and bargaining tool, labor protest in parts of the Anthracite Region took a different approach in the 1930's. John Bodnar's "The Family, Economy, and Labor Protest in Industrial America" traces miners' efforts to deal with a declining market for anthracite and reduced job opportunities and the effect these had on their families. Bodnar sees the formation of the separate United Anthracite Miners as having two goals: to equalize the production of coal among the miners and among the collieries to preserve the community, and to replace the John L. Lewis dominated United Mine Workers of America with an organization more sympathetic to their needs. This program was based on a local ethnic culture which believed in strong familial ties and community support. Bodnar sees another dimension to the problems associated with economic decline and labor dispute—a higher incidence of depression and schizophrenia among miners and their wives. Bodnar concludes that changing family patterns and increased health problems underscore the responsibility mining families felt toward their community during periods of strike and economic depression.

All of these studies underline the inter-relationship between ethnicity and labor in Pennsylvania's Anthracite Region. They provide insights into both the conflicts and the efforts at cooperation which were a part of everyday life in the Region. While not the definitive work on the subject, each essay provides new insights in a particular area of regional study. Together, the essays provide examples of the range of resource materials available on Pennsylvania's ethnic and labor history and illustrate the value of an interdisciplinary approach to the Region's history. Finally, as with all historical studies, they raise new questions and create opportunities for additional research and writing.

The symposium on "Ethnicity and Labor in the Anthracite Region" and the publication of HARD COAL, HARD TIMES was a joint effort of the Scranton Anthracite Museum Associates and the Pennsylvania Historical and Museum Commission and were made possible by a grant from the Pennsylvania Humanities Council, a Statewide Funding Agency funded in part by the National Endowment for the Humanities. Special thanks for assistance in preparing the symposium and publication are due Craig Eisendrath, Pennsylvania Humanities Council; Dr. Elizabeth Jewell, Scranton Anthracite Museum; Mary Ann Landis, Eckley Miners' Village; Mrs. Adrienne Horger, President, Scranton Anthracite Museum Associates; Harold Myers, Division of Archives and History; Jeanette Harvey, typist; and the participants whose work appears in this publication.

David L. Salay
Director
Anthracite Museum Complex

Chapter One

THE MINERS OF ST. CLAIR:
FAMILY, CLASS, AND ETHNICITY IN A MINING TOWN IN SCHUYLKILL COUNTY 1850-1880

Anthony F. C. Wallace

It is difficult to avoid approaching the subject of ethnicity and labor in the coal regions of Pennsylvania without taking ethnic diversity and social conflict as the motif. After all, the successive layers of immigration—the Pennsylvania German farmers in the eighteenth century, the English, Welsh, Scottish, and Irish miners and entrepreneurs in the early nineteenth century, then a new wave of German miners during the Civil War, and finally the various Slavic nationalities from the 1890s on—have indeed left an ethnic kaleidoscope of languages, accents, cuisines, costumes, and religious denominations. And the sensational trials, and executions of the Molly Maguires in the 1870's, and the bitter strikes and anti-labor violence then and subsequently have cast a long shadow, showing in high relief the divisions between classes and between ethnic groups.

In this paper, however, I propose to begin the consideration of ethnicity from the other direction. I shall suggest that in the period under consideration, 1850 to 1880, there was in fact, to an important degree, a common culture shared by all the ethnic groups in the region, and that heightened consciousness of ethnic differences, and the venting of ethnic antagonisms between strikers and strikebreakers, was artificially induced in the course of social conflicts that had their source in economic rather than ethnic competition.

My views on this problem have developed in the course of an on-going study of the town of St. Clair, and most of my examples will be drawn from my St. Clair materials. But processes of change and conflict at work in St. Clair were, I believe, experienced throughout the anthracite district, and certainly in most of Schuylkill County.

THE "ANTHRACITE CULTURE"

One cannot approach the historical ethnography (or social history) of the anthracite district without recognizing that it was a distinct social entity based on the mining and export of coal, with its own internal structure and external connections (markets and politics). In this sense it was a community, a society, almost a state within a state, with its own unifying culture.[1]

Even so, it is perhaps risky to propose that in an area with a newly developing industry and multiple layers of ethnically distinct recent immigrants, a common culture could exist. And I am, in fact, one of those anthropologists who generally emphasize diversity as much as sharing. But I am also one of those anthropologists who do not restrict the concept to symbol systems—that is to belief and ideology—but rather see a culture as a set of techniques or, more precisely, rules for making decisions as to how to solve recurrent human problems. Many of these techniques are widely shared, or are at least available for general use in a population of interdependent individuals. The concept thus embraces technology, language and communication, social organization, and belief and ideology (religious, political, and economic). Furthermore, the anthropologist no longer views culture merely as a patchwork of "traits" or "attributes" distinguishing tribes, nations, or culture areas from one another, but considers that technology, social and economic organization, and religious belief and ritual will have systematic ("functional") relationships. Thus stress, or change, in one domain of culture will have effects on others.

From this point of view, the significance of ethnicity and class is political rather than culture-wide. Ethnicity, as an attribute recognizable in most historical data, would seem to have two separable components: national origins, and religion. As we shall see, social class cuts across both dimensions of ethnicity, although there are strong correlations. In a broad way, one may view the anthracite community as a two-class society—the workers and the capitalists.[2] But this is far too simplistic an analytic scheme to be useful in understanding the social conflicts of the period. The *Miners' Journal* editorialized about the "social classes" in a simple way, to be sure, and the miners' union on the one hand, and the spokesman of the land owners and mine operators, on the other, spoke as if the two classes were as clearly defined as the two political parties (which were, of course, in reality both moved by a variety of special interests). But for many persons, class membership was an ambiguous and fluctuating condition that cannot be easily reduced to occupational classification or the ownership or nonownership of the means of production. A miner today might, by borrowing from kinsmen, tomorrow acquire an interest in a mine and become a capitalist, only to lose the property a few months later. Some injured miners became merchants and tavern keepers; the wives of others opened dress shops and beer parlors. And, within the managerial class, there existed a fundamental conflict of economic interest between the great transportation and coal-and-iron corporations (in Schuylkill County, the Philadelphia and Reading) and the small mine operators and landowners, many of whom were extended family businesses.

Thus, within the anthracite community, ethnicity (i.e., natural origin and religion) and class provided lines of *potential* cleavage and alliance. But—as I shall argue—ethnicity did not assume an important role until after economic and class conflicts had become intense. The reasons for taking this position are as follows. First of all, it is important to remember that all the major ethnic groups in Schuylkill County in 1850 had either originated in the British Isles or (in the case of the Pennsylvania Germans) had lived for generations in an English-speaking country, the United States. Some Pennsylvania Germans spoke German primarily and English only in dialect, but few of them worked in the mines. The miners—English, Welsh, Scots, and Irish—all spoke English (although some Welsh insisted on maintaining Welsh newspapers and other bilingual usages, to the annoyance of their neighbors).

Second, the common culture included a virtually universal technology and material culture. Central to the economy was the standard British coal mining technology, knowledge of which was brought over by experienced English, Welsh, and Scottish miners, and which was supplemented by imported English manuals on such subjects as mine ventilation, plan of working, method of sinking, pumping water, etc. The Irish were at a disadvantage here. Although there was coal in Ireland, it had never been extensively developed and there were few if any immigrants coming directly from Ireland who could present themselves as "miners" proper. (A sizable minority did, however, move to England before coming to America.) The majority thus had initially to accept work, under worse conditions and slightly lower pay, as "laborers"— that is as miners' helpers. But after acquiring the requisite knowledge and information, Irish laborers advanced to the rank of miner too.

Third, members of all ethnic groups soon learned the map of the regional organization, that is, the physical layout of canals, roads, and railroads that joined the miner, the mine patches, the mining towns, the regional centers, and the county seat; and the corresponding economic and political network that managed the whole system and connected it with markets and competing coal regions and centers of money and political power, in Philadelphia and New York, in Mauch Chunk and Wilkes-Barre, in Harrisburg and Washington. They all traveled the same roads, rode the same trains, voted in the same elections. They all knew the social roles and ranks of miners, laborers, engineers, lawyers, mine operators, coal agents, land owners, and railroad and canal officials. They all recognized that there were great inequalities of wealth and power, and, being British in origin, undoubtedly saw a clear-cut class structure. But they also undoubtedly believed that some working people, or their children, would climb the ladder of success.

Finally, all shared fundamentally the same kinship system in ideology and in practice (which made possible a small but significant number of inter-ethnic marriages). Such differences as may have existed were slight in comparison with the contrast between these Western European forms and the kinship systems of, let us say, the American Indians and many other non-European peoples.[3] Minimally, the system emphasized monogamous marriage, a residence pattern in which a married couple with children ordinarily occupied their own household, with perhaps a few ailing or single relatives and paying boarders, and the maintenance over time of an extended family network. The nuclear family household seems so central to Americans, in part because the U.S. census is constructed as a household census, and this fact tends to make difficult the recovery of information about extended families from census schedules (except in the 1880 census). But from other sources—letters, diaries, church records, company books, newspapers, and so on—it is clear that for all social classes, the bilateral extended family, reckoning marriage and descent through both males and females, was in this period crucially important in providing security for the ill, injured, aged, widowed, orphaned, or impoverished, and as a network for the mobilization of capital for joint ventures, such as the purchase of real estate or investment in a business.

Let us examine the concept of "family" more carefully. Conventional English terminology employs the word "family" as a gloss for a wide variety of kinship concepts and arrangements which should for analytical purposes be distinguished explicitly. In the U.S. census, a "family" consists of the members of some kinship group (usually but not always a "nuclear family" of husband, wife, and children) residing under one roof, with a person denominated as "head of household" to whom the others are related by descent and/or marriage or adoption (in the 1880 census these relationships are explicitly stated). But for a given person (in anthropological jargon, an "ego"), the boundaries of economically and emotionally important kindred usually include members of other households (census "families") as well. These other members of ego's kindred may include grandparents, brothers, and sisters, uncles and aunts, cousins, sons and daughters, nephews and nieces, and grandchildren, and their spouses and *their* children, and congeries of inlaws. Reckoning is done bilaterally, up, down and sideways, so as to include relatives through both males and females, and by both descent and marriage. An extended family may have several nuclear family units in the same town and also additional members in other towns, near or distant. Where such an extended bilateral descent-group coalesces for generations around an estate of some kind—whether land, house, mine, or manufactory—it may be termed a cognatic descent group, in anthropological jargon.

In the nineteenth century, when many nuclear families provided large numbers of children, such a descent group could include dozens of adult individuals, to whom members could appeal for support and assistance in various ways

—money, transportation, temporary housing and food, assistance in obtaining employment, nursing care, to name a few. For working people, this network of kin was the primary social safety net; for those sick or unemployed who had no kin to help them, there was only the poorhouse. Outside charity by church congregations and middle-class philanthropy, and the assistance of unions and workingmen's benevolent associations were a temporary supplement when hard times overwhelmed the kinship network.[4]

There were, however, important differences among—and within—the ethnic groups in religion and, probably, in values with respect to authority. The obvious fact that the Catholic Irish were the object of suspicion as "Papists" by many Protestants should not distract attention from the fact that there also was a range of Protestant orders from the relatively high-church Espiscopal, Lutheran, and Presbyterian to the various non-conformist, low-church denominations—Baptist and Methodist, principally. But all of these were also split by national origin, so that there were Irish Catholics and German Catholics, Welsh and Scottish Presbyterians, Welsh and English Methodists.

Most of the national denominations got along with one another reasonably well. It was only the Irish Catholics who were singled out for public attack as the 1850s wore on, for reasons that I shall endeavor to explain. Religious and ideological diversity is after all a common human condition, not only in Western states but in cultural settings widely divergent in many respects from European traditions. And in this part of America, among the middle upper classes, national origin and religious preference were not the primary qualifications for membership. Thus, in the anthracite domain, one may note that the principal actors were of extremely diverse social origins. Franklin Benjamin Gowen, leader of the Reading coal, iron, and railroad combination, was born of Irish parents; the Careys and the Bairds, who owned the coal rights under the town of St. Clair, also were born of Irish parents and were strong Irish nationalists; Benjamin Bannan, the editor of the *Miners' Journal,* was Welsh, and so was William H. Johns, the most (indeed, the only) successful coal operator at St. Clair. The lands around St. Clair (including those mined by Johns) were owned by members of the Wetherill and Seitzinger families, respectively English Quaker and Pennsylvania German in origin. And the Careys' and Bairds' some-time partners in business and land ownership were Isaac Lea, of English Quaker stock, and Abraham Hart, a leading Jewish businessman of Philadelphia and New York. But, despite the diversity of national origins of this group of managers whose decisions affected the people of St. Clair, not one was Catholic. And it was from this group—the owners and operators of the mines—that the attacks upon Catholicism emerged, not from Protestant workers.

Indeed, there is no indication among the working people, of St. Clair at least, that there was any particular concern about religious differences. What did at times divide ethnic groups along lines of national origin (or race) was an economic issue—strikebreaking. The information is not easy to gather, but from the emigrant letters of Welsh miners in the anthracite district, and from union leader John Siney's speeches and correspondence, it is clear that in prolonged strikes, mine operators imported strikebreakers of alien ethnicity from outside the district, or even from Europe, in the belief that they would be more willing to cross picket lines. Thus one Welsh miner, writing from Pottstown in 1865, told of meeting an old friend from St. Clair, who reported that the operators had pooled together to import a whole company of Welsh miners to break a strike; and the same man, six years later, complained that to break a strike the masters "got a number of old spineless Irish to be blacklegs (turncoats) at one of the pits and succeeded in getting hundreds of soldiers here to guard them. . . . Had it not been for those old Irish we would have won a complete victory two months sooner."[5] Later, in the 1870s, John Siney complained bitterly that the operators were importing inexperienced southern blacks to break a strike by English and Welsh miners in Illinois.[6]

Clearly, Schuylkill County was not a peaceable kingdom in the 1850s and 1860s. To the contrary, it was a cockpit of conflict. But the conflict was not essentially over ethnic or cultural or even religious issues but, as I shall argue in the last section, over who should survive and what kind of society should emerge from the ruins of the coal trade, as the Schuylkill County system of small entrepreneurs collapsed in the face of technological and economic challenges which it was not equipped to handle.

ETHNICITY, OCCUPATION, AND FAMILY IN ST. CLAIR, 1850-1880

It is appropriate to use the manuscript schedules of the 1850 population census as a baseline for ethnicity and occupation statistics in St. Clair.[7] The enumerators visited the town just before it was officially organized as a borough, later in that same year, so we have to rely on the figures for the part of East Norwegian township north of Port Carbon. Since most of the population of the township was concentrated at St. Clair and the mine patch called Mill Creek, about a mile below St. Clair proper, I see no difficulty in comparing the 1850 figures with later censuses of St. Clair proper.

The total population of East Norwegian in 1850 (leaving out Port Carbon, which was listed separately) was 1,031, of whom 535 were white males, 495 white females, and one colored male. The enumerators counted 193 "families" occupying 194 dwelling houses. The occupations listed for males sixteen years and older were primarily associated with mining and its supportive trades. For agriculture, only three farmers were listed in the entire township. The bulk of the employed male population were listed as "miner" and "laborer." The term laborer, I assume, is used to refer to mine laborer (i.e., the contract miner's underground helper) and to other miscellaneous underground jobs, such as mule driver, and to some above-ground jobs as well, such as ditch digger. (Later censuses distinguish between "laborer" and "laborer in the mines.") Lumping all laborers and miners together as underground mine workers, the total number of underground mine workers was on the order of 193, of whom 100 were listed as "miners" and ninety-three as "laborers". The relation between ethnicity and occupation is clear. Of the 100 miners proper, seventy-eight were British-born (i.e., from England, Scotland, or Wales) and only sixteen were Irish-born; the remainder were German-born (two), and Pennsylvania-born (six). Of the laborers, only eight were British-born, and sixty-eight were Irish; the remainder again were from Germany (three), Pennsylvania (thirteen), and elsewhere in the U.S. (one).

By 1870, the relationship between ethnicity and occupation had changed considerably. The town of St. Clair (now a separately enumerated borough) had more inhabitants than all of Norwegian in 1850, 5,726 *in toto* (1,126 families, 1,156 residences), 2,943 white males and 2,783 white females. The foreign-born still constituted a large proportion of the total, 2,282 in all—almost exactly forty per cent. The English, Welsh, and the German-born now dominated the "coal-miner" category, but there were now proportionately more Irish-born miners. In a two-thirds sample, the figures were Irish thirty-five, English fifty-six, Welsh forty-one, German twenty-three, and U.S.-born nine. And in the "laborer in mines" and "mule driver" categories, the Irish-born were positively outnumbered by a combination of British, Germans, and the Pennsylvania-born (although there were still few British-born laborers). The figures were: Irish fifty-one, English twenty-two, Welsh twenty-one, German fifteen, and U.S.-born fifty-four.

One cannot avoid the conclusion that in the first two decades of deep mining at St. Clair, the occupational disadvantages of Irish mine workers were ameliorating. No doubt the reason was that after some experience underground, and with time to raise the cash, some of the Irish laborers were able to accumulate the skills, the tools, and the credit to set up as contract miners themselves. By 1870, furthermore, the Irish in St. Clair (and elsewhere in Schuylkill County) were prominent as storekeepers and publicans, and were in the professions. Furthermore, in the anthracite region at large, the Irish vote had become an important factor and Irishmen were representing the counties in the legislature in Harrisburg. There is little to suggest that the ethnic cleavage between the British and the Irish was sufficiently important, among working-class people at least, to prevent working side by side in the mines, or to produce any marked residential segregation among the three wards, or even to stop a significant degree of cross-ethnic marriages. In fact, about ten per cent of the working-class households contained married couples of different national origins (English-Irish, English-Welsh, Welsh-Irish, Scottish-English, or north-and-south German). In view of the fact that any married couple represented a link in an extended family network, it is reasonable to infer that far more than ten per cent of the extended family groups in the town were, in fact, multi-ethnic.

Let us now look a bit more closely at the extended family system as it operated in St. Clair by examining the census and other records of three extended family connections: the Phillips/Williams group, which was at least mostly Welsh; the Carroll/Kinney/Jordan group, which was Irish; and the Tempests, who were English, Irish, Welsh, and Scottish. In no case do we have an even reasonably complete genealogy; what we observe is only a fraction of the whole. But we can see enough to infer something of how the extended family system worked as a social-support network.

We may begin with Thomas Phillips, an "enterprising, moneymaking Welshman" who emigrated to the United States sometime before 1864. He may have been the same Thomas Phillips who lived in St. Clair in 1862, worked in a mine at Wadesville, a mile away, and wrote an informative letter back home that has fortunately been reprinted in a collection of Welsh immigrant letters.[8] In any case, one Thomas Phillips acquired an interest in a colliery at Summit Hill, in Carbon County, and became superintendent there. In 1864 he "succeeded to" a general store at Hyde Park, near Scranton, and appointed as manager a man named John Williams. By 1876 Phillips was worth about $100,000. In that year, back at St. Clair, he came to the rescue of "a relative" named Daniel Williams, whose general store had failed. Phillips bought the stock at a sheriff's sale and transferred it to Christopher, Daniel's son, aged about twenty-three, who had been a clerk in his father's store. The credit rating agency that reported on the transaction declared the new business of Christopher Williams "will be safe with the endorsement of Thomas Phillips." Daniel Williams, his wife, and children (eventually, at least nine of them) had been living in St. Clair since 1860 at latest, when the census reported him as a merchant. He had been born in Wales and emigrated to the United States about 1842.[10] Evidently, in light of the fact that Phillips was aiding a relative named Williams, the connection was established through at least one affinal link; that is, a Phillips had married a Williams, and the connection was traced through a woman.

The Irish case centers in a coal miner named Timothy Carroll, who, in 1870, lived in St. Clair as a widower, age thirty-six, with a son John, age twelve, a daughter Ellen, ten, and a son James, eight. Also living in the same household was Margaret Jordan, his aunt, age fifty-nine. She and her husband, James, a mine laborer, had been listed in the 1860 St. Clair census. Next door to Timothy Carroll lived his brother, Dennis Carroll, and his new wife, Cecelia; and elsewhere in town lived their (probable) sister, Mary, and her husband,

Michael Kinney, a "huckster" in New York. In 1874, Timothy Carroll worked in the Wadesville Shaft (where many Irishmen worked) and was a principal leader in Siney's union.[11] Between 1870 and 1880, a number of things happened to this extended family: Mary and her huckster husband died or left St. Clair (leaving their two children behind); Dennis' wife, Cecelia, deserted him or died, leaving him without children; and their father, Owen Carroll, came to live in St. Clair. In 1880, Timothy and Dennis still lived in adjoining houses. Timothy's sons had left home, and the household now consisted of Timothy, still a coal miner; his daughter Ellen, nineteen, housekeeper; his father, eighty; his aunt, sixty-nine; and a nephew, Michael Kinney, four. Michael Kinney's sister, Mary (seven), lived next door with her uncle Dennis.[12] The last we hear of Timothy (if it was the same Timothy) is that a Tim Carroll suffered a broken rib in Beechwood Colliery, at nearby Mt. Laffee, in 1881.[13]

Our third case is a complex cross-ethnic extended family involving English, Scots, Irish, and Welsh over at least three generations. The presence of the Tempest/Stephenson connection in St. Clair first becomes visible in the 1850 census, which lists a Joseph Tempest, miner, age thirty-eight from England, as head of a household including his wife, Catherine, from Scotland, also thirty-eight, their five boys, all born in England except the last (age one year), plus a twelve-year-old girl with a German name who was probably a servant, and a boarder, a miner from England. By 1860, after siring a daughter, Joseph had died and the widow Tempest was keeping a boarding house for coal miners, which she owned in the south end of town. Her oldest son, Thomas, had gone off to Australia about 1859 and married an Irishwoman named Bridget Mack. The rest of Mrs. Tempest's children, with the exception of the oldest, were living with her, the two older boys mining coal. She employed a Welsh servant girl; and, about 1861, a second son, Joseph, had married a Welsh girl. By 1866 or 1867, Thomas had returned with his Irish wife and four children and, in 1868, joined John Siney and twelve other miners of varied ethnic backgrounds to found the highly successful Workingmen's Benevolent Association.[14] In 1870, the family was considerably dispersed. The widow Catherine, now fifty-nine, had gone to live with an English coal miner, a forty-eight year-old widower named William Stephenson, who occupied a house across the street from her boarding house, along with his handicapped son, John, a mine laborer, Catherine's own daughter Hannah, sixteen and now a school teacher, and a fifty-nine-year-old woman from Scotland named Elizabeth Phillips. Catherine's grandchildren were also grandchildren of William Stephenson, and it is tempting to speculate that the Scottish washerwoman was also her relative.

During the 1870s, the Tempest brothers all had their own households. They appeared in violent scenes. Four of the brothers were miners and one was the town constable. In October of 1870, Joseph and five other men were badly burned in a gas explosion in the Wadesville shaft.[15] And in 1874, James was recommended to the Pinkerton spy in St. Clair (assigned to watch the W. B. A. for signs of terrorist activity) as a suitable "butty" when the spy began looking for work. The spy started to work as a laborer with James in the Wadesville shaft, but was soon thrown out of work by James's going on a spree. They started again a week later but James was injured by a fall of coal ("he had a very narrow escape from death") and the job lapsed again. The spy was finally thrown out of his job by James on account of James's going to work with his brother. "Work is scarce now owing to partial suspension of the P. and R. mines," reported the spy.[16] The reader will note that loyalty to kin took priority over loyalty to an unrelated partner even though both were members of the W. B. A.

By 1880, all the Tempests had left St. Clair except the aging widow Catherine, who by now had married William Stephenson, and James's daughter Isabella, age nine, who lived with her grandmother and her grandfather. William Stephenson, now fifty-nine, was still mining coal. The rest of the Tempests had moved to Shenandoah, a larger mining town about eight miles to the north. By then, one of the brothers, Martin, the constable, and one of the wives, James's Maria, had died, and James had married his widowed sister-in-law, Matilda, and he and his sons lived with her and her children. Brothers Thomas, Joseph, and Andrew had large families of seven or eight children. All four brothers still were coal miners and all the boys over ten years of age worked as slate pickers, mule drivers, or laborers.[17]

From examples such as these, and from other sources of information as well, particularly collections of immigrant letters, one must conclude that in some respects the extended family connection was more important than the nuclear or household family among working-class people in this period.[18] In particular, it provided the individual with the only reliable social safety net that society provided, short of the poorhouse. There was no unemployment compensation, no social security, no free medical care. As we have seen, the extended family connection, reckoning through both descent and marriage, entitled a member to claim residence when widowed or orphaned, financial aid when bankrupt or injured, a job when out of work. The network of extended families in the town of St. Clair (and in other industrial towns) provided what was in effect an emergency support fund of housing, money, and services that I would argue, was crucial to the stability or even survival of a large part of the laboring population.[19]

ETHNICITY AND SOCIAL CONFLICT

During the two decades from 1850 to 1870, the burden on what may be called the working people's emergency fund probably increased. Three main reasons may be advanced for such a putative increase (apart from the business cycles of the coal trade and the stresses induced by the Civil War). First of all, colliery operators deliberately encouraged the over-migration of extended families for the purpose of providing an over-supply of workers, and, in turn, for keeping down wages (or at least so the well-informed author of *White Slaves of the Monoplies* charged in 1884).[20]

Second, the high rate of colliery failures meant chronic unemployment for many workers. Clifton Yearley, the economic historian of the Schuylkill coal trade to 1870, identified 568 separate mining firms (some of them, of course, successively operating the same colliery) and found that fewer than half "survived more than one year of business"; only six per cent lasted fifteen years.[21] The reason for this extraordinary rate of failure, I have argued elsewhere, was not so much the seasonal and cyclical ups and downs of the market for anthracite as the vulnerability of inadequately capitalized and technologically unsophisticated collieries to physical disaster. Crushes, gas explosions, underground mine fires, and flooding (to put out the fires) put many collieries out of business repeatedly for weeks, months, or even years at a time. Let us look briefly at one case from St. Clair—the famous, or infamous, St. Clair Shaft. This colliery opened with much fanfare in 1854, touted as the first vertical-shaft colliery in the region, 450 feet deep from grass to gangway. It closed down permanently twenty years later after producing only about five hundred thousand tons of coal (profit on which, at the normal ten cents per ton above operating expenses, would amount to less than half the capital originally invested). During its twenty years, it was operated by six owners (the last, the Philadelphia and Reading Coal and Iron Company). During that time, in addition to dozens of small fire-damp explosions in individual breasts that killed and burned miners in groups of a few at a time, there were two mine fires (one of them a duplicate of the Avondale fire, but without casualties because it occured at night when no one was in the mine) followed by floods, and two breaker burnings. The fate of the St. Clair Shaft is not atypical. Of the seven deep mines under and around St. Clair in the period 1850–1880, at least four, and possibly six, closed down permanently as a consequence of a similar sequence of disasters.[22]

The third reason for the increasing burden on the extended family support system is related to the second. It is the increasingly high incidence of underground accidents, involving a single person or at most a half dozen. Most were caused by falls of stone or coal, but a large number were the result of explosions of fire-damp (methane) collecting at the top of the steeply pitching breasts as a result of inadequate ventilation. Such gas explosions were inevitable. In the St. Clair shaft (according to the Pinkerton agents' reports), men were expected to work in breasts with four feet of the lighter-than-air fire-damp reaching down from the ceiling.[23] Let us look at the figures from August 1, 1870 to July 31, 1871 (the year following the census-enumeration, and also after work resumed following the W. B. A. strike, and two years after the passage of the first mine safety law in Pennsylvania). During this year, there were nine fatal accidents (two Welshmen, three Englishmen, and four Irishmen) and twenty-four non-fatal accidents serious enough to be published in the annual mine inspector's report. Most were caused by gas explosions (five fatal and sixteen non-fatal) and falls of coal (three fatal and three non-fatal). Most (sixteen) occured in two fiery mines, the afore-mentioned St. Clair Shaft and the Pine Forest Shaft, where approximately eleven per cent of the inside work force were killed or injured.[24] The overall 1870–1871 casualty rate for inside miners and mine laborers listed in the 1870 census was about five per cent.

In effect, one can say that deep-shaft mining in the steeply pitching veins of Schuylkill County was a gradually increasing disaster that reached a critical condition by the late 1850s and early 1860s. The industry was temporarily salvaged by the need for anthracite during the Civil War, but it virtually collapsed between 1865 and the early 1870s, when the Gowen and the Reading interests took over, in the process eliminating the sub-class of small colliery operators and landowners. During this time of travail, the Workingmen's Benevolent Association of St. Clair was organized to supplement the overburdened emergency resources of working-class families and to represent the miners' interest as a class. Despite Gowen's efforts to associate it with the Irish Catholic image of the Molly Maguires, the W. B. A. was not an ethnic organization, ethnic and regional divisions were both explicitly recognized as enemies of union solidarity. Partly as a result of its lobbying effort, the Legislature was persuaded in 1869 to pass an Act providing safety regulations for Schuylkill County coal mines, particularly as regarded ventilation; after the Avondale disaster, this act was improved and extended to cover all the anthracite counties. And, of course, the W. B. A. negotiated the famous "sliding scale" of wages with the operators in an effort to stabilize the industry financially.[25]

Ethnicity was not a factor in the structure or purposes of the W. B. A.; occupational, or class, solidarity could only be weakened by ethnocentric divisions. Ethnicity was, however, deliberately injected into the issue by the threatened operators and their spokesmen, from Benjamin Bannan of the *Miners' Journal*[26] to Franklin Benjamin Gowen, the devious impressario of the Philadelphia and Reading Railroad and Coal and Iron Company.[27] The miners had traditionally been blamed by colliery operators for all accidents and disasters, attributing them to carelessness if not malice. Such a position, of course, protected the operators from a potentially uncomfortable awareness of the technological backwardness, inefficiency, and hazardousness of their underground workings. As the economic pressure, first on the operators and then on the Coal and Iron Company, increased through the 1860s and 1870s, a virtual mythology was constructed by representatives of the coal trade. This mythology blamed the trade's difficulty not on an inadequate technology (a technology, indeed, that perhaps could never have been adequate at that date to deal with the geology of the region), but on criminal elements among the working class. This criminal element was increasingly identified by the name of "Molly Maguires." Thus, in writing to Henry Carey in July 1868, reporting that the mines at St. Clair had been stopped by the "eight-hour raiders," Carey's agent, Henry Russell of Pottsville, referred to the "raiders" as "the scum of the Molly Maguires."[28] Actually the strike, which spread all across the anthracite district (including, of course, miners of all ethnic varieties) and lasted until August 28, had been called in protest against the colliery operators' failure to implement the new law mandating an eight-hour workday without reducing wages. It was organized and directed by the W. B. A. In 1870, Gowen publicly charged that the Molly Maguires were the terrorist arm of the W. B. A., enforcing their edicts by force and threats of force.[29] And the Pinkerton investigation, ordered by Gowen and supplemented by the Coal and Iron Police, resulting in the highly publicized trials and hangings of twenty so-called Mollies, was (whatever the personal guilt of individuals) an elaborate drama that communicated throughout the coal regions (indeed, the nation) the old contention that an Irish Catholic conspiracy existed aiming to subvert the state. And later on, of course, social problems in the anthracite district were attributed to the presence of unassimilated, if not unassimilable Italian and Slavic workers imported as strikebreakers during the generation of labor conflict after the "long strike" in 1875.

This is not the place to deal again in detail with the Molly Maguires issue. It is necessary, however, to point out that the Molly Maguire publicity contributed to a perception on the part of the middle-class reading public that the anthracite mining district was a setting of social anarchy. The sociologist Peter Roberts, trained in the social Darwinist tradition of William Sumner at Yale, in 1904 saw the anthracite districts, with their twenty-six ethnic groups, as the victim of both "the Malthusian law" and "Gresham's law." The "elite" of the Anglo-Saxon and German miners improved themselves and left the coal regions; the "degenerate" dregs of these "races" remained behind, mixed with an increasing horde of "needy, ignorant and incapable" Slavic and Italian immigrants. Thus a kind of reverse evolution occurred:

> The character of the population of this area has perceptibly deteriorated in the last thirty years. A selection has been effected but in a retrogressive sense.[30]

Such a view is still expressed in scholarly journals; see, for instance, Rowland Berthoff's essay, "The Social Order in the Anthracite Region, 1825–1902," published in 1965, which saw no "encompassing order," no system of "reciprocal rights and duties" in the whole region.[31]

Such views of ethnicity as a critically divisive feature of the anthracite culture, it appears to me, have been an artifact of a situation of social conflict. To the hardpressed small operators of the 1860s and 1870s, ground between the wage and safety demands of workers and the ambitions of the transportation and coal and iron corporations, and soon to be swallowed by the Reading, it was easy to blame their troubles on careless, rowdy, hostile miners. It was easy to imagine a secret, underground Irish Catholic terrorist organization (after all, a generation before a similar Protestant political effort had been directed against the Masons), even though the "terrorists" seem to have been a set of Irish and Welsh street gangs. And, ten years later, after the Reading took over, it was easy to believe that it was the Reading's plot to drive out the respectable Irish, German, English, and Welsh that led to the importation of the careless, rowdy, hostile Italians and Slavs. All of this covered over two uncomfortable facts: the ability of the miners to organize cross-ethnic unions and the inability of the operators (including the Philadelphia and Reading Coal and Iron Company) to find effective technological solutions to the problems of deep-shaft mining in steeply pitching seams. These views, furthermore, it was no doubt hoped, would tilt middle-class public opinion away from sympathy for the miners and toward legislative and administrative actions in favor of the mine operators. The victims, of course, were not only the miners—especially the Irish—but also the mine operators themselves.

CONCLUSION

In this paper I have argued that, in Schuylkill County anthracite mining communities like St. Clair, in the period 1850–1880, Irish Catholic versus British Protestant ethnicity did not represent a major social or cultural problem. Many important cultural features were, in fact, shared, from mining technology itself to the extended family system, which served as the major social-support network and united ethnic groups by marriage. Even though ethnic antagonisms were aroused when ethnically alien strikebreakers were imported by mine operators, such divisions did not prevent the formation of effective cross-ethnic miners' unions. Such social problems as drunkenness, crimes of violence against the person, and industrial sabotage were not confined to any one ethnic group. But an image of the coal region, and particularly of Schuylkill County, as increasingly a patchwork of Catholic ethnic enclaves, nourishing incompetent miners and anti-Protestant terrorists was deliberately created by representatives of financially hardpressed mine operators, including the coal and iron companies, in order to provide a face-saving explanation for the failure of the deep-shaft mining system in this period and to bring public pressure to bear on recalcitrant miners' unions.

NOTES

[1] William Gudelunas' thoughtful paper on the "Irish factor" uses the term "internal colony" to describe the status of Schuylkill County in the period after Gowen had acquired control over the land, the collieries and furnaces, the police, and to a degree the governmental institutions affecting the county. Aurand quotes Francis Nichols as asserting in 1902, in a similar vein, that the anthracite counties, bound together by the interests of capital and labor, constituted "a sort of separate and distinct state, called by its inhabitants 'Anthracite.'" See William Gudelunas, "The Rise of the Irish Factor in Anthracite Politics, 1850–1880," Pennsylvania Historical Association, October 1981; Mark G. Hirsh, "Class Ethnicity and the Irish Miners of the Lower Anthracite Region of Pennsylvania, 1850–1880," Pennsylvania Historical Association, October 1981; Harold W. Aurand, *From the Molly Maguires to the United Mine Workers: The Social Ecology of an Industrial Union, 1869–1897* (Philadelphia, 1971).

[2] Katz employs a two-class model in analyzing nineteenth century Buffalo, N.Y. and Hamilton, Ontario. The classes are determined by their relationship to the mode of production (i.e. ownership or labor); the attributes of a class system are: (1) the classes are ordered hierarchically; (2) class interests are permanent; (3) members share a sense of class identity, ranging from passive "awareness" to militant "consciousness"; (4) the classes are socially isolated from each other. See Michael B. Katz, Michael J. Doucet, and Mark J. Stern, *The Social Organization of Early Industrial Capitalism* (Cambridge, Mass., 1982).

[3] For an overview of anthropological models of kinship, see Robin Fox, *Kinship and Marriage* (Middlesex, England, 1971).

[4] For more extended discussions of the importance of the extended family in nineteenth century industrial communities, see Anthony F. C. Wallace, *The Social Context of Innovation* (Princeton, 1982) and "Extended Family and the Role of Women in Early Industrial Societies," *Working Papers from the Regional Economic History Research Center* (Wilmington, Del.), Vol. 5, nos. 2 & 3, 1982.

[5] Alan Conway, ed., *The Welsh in America* (Minneapolis, 1961), 175, 191–192.

[6] Edward Pinkowski, *John Siney, the Miners' Martyr* (Philadelphia, 1963), 187.

[7] The original schedules for the 1850, 1860, 1870, and 1880 censuses for East Norwegian Township (1850) and St. Clair borough (1860, 1870, and 1880) are located in the National Archives, Washington, D. C. I worked with copies at Eleutherian Mills Historical Library, Wilmington, Delaware, and the Federal Records Center, Philadelphia, Pennsylvania.

[8] Conway, *The Welsh in America*, 167–170.

[9] Middle States Reports, *Middle States Reports for 1876* (New York). Copy at Eleutherian Mills Historical Library.

[10] U.S. Census, Population Census Schedules for Borough of St. Clair, 1860, 1870, 1880.

[11] "Molly Maguire Papers," Society Collection, Historical Society of Pennsylvania, Philadelphia, Pennsylvania. Report of Pinkerton Agent P. M. Cummings, 6 April 1874.

[12] U. S. Census, Borough of St. Clair, 1860, 1870, 1880.

[13] Pennsylvania Bureau of Mines, *Reports of the Inspectors of Coal Mines of the Anthracite Coal Regions of Pennsylvania for the Year 1871* (Harrisburg, Pa, 1872), 36.

[14] Pinkowski, *John Siney,* 15.

[15] Pennsylvania Bureau of Mines, *Reports of Mine Inspectors,* 1870 (Harrisburg, 1871), 22.

[16] "Molly Maguire Papers," Society Collection, Historical Society of Pennsylvania, Cummings reports on 10 March, 21 March, 1 April, and 14 April 1874.

[17] U.S. Census, Borough of St. Clair, 1860, 1870, 1880. Borough of Shenandoah, 1880.

[18] In addition to Conway, *The Welsh in America,* relevant collections include: Charlotte Erickson, ed., *Invisible Immigrants: The Adaptation of English and Scottish Immigrants in Nineteenth Century America* (Coral Gables, Fla. 1972); Thomas W. Leavitt, ed., *The Hollingworth Letters: Technical Change in the Textile Industry, 1826–1837* (Cambridge, Mass., 1969).

[19] See also Wallace, *The Social Context of Innovation* and "Extended Family and the Role of Women."

[20] (John Fitzpatrick), *The White Slaves of Monopolies* (Harrisburg, 1884).

[21] C. K. Yearly, Jr., *Enterprise and Anthracite: Economics and Democracy in Schuylkill County, 1820–1875* (Baltimore, 1961), 58–59.

[22] See Wallace, in press, for more detail on the frequency of accidents and disasters and their economic impact. Wallace, "The Ventilation of Coal Mines: A Nineteenth Century Case of Social Conflict and Technological Failure," Canadian Historical Association, June 1982 (forthcoming).

[23] "Molly Maguire Papers," Society Collection, Historical Society of Pennsylvania. Cummings report for 6 April 1874.

[24] Pennsylvania Bureau of Mines, *Reports of the Mine Inspectors,* 1870, 1871.

[25] General histories of the W. B. A. are given in Pinkowski, *John Siney,* and Charles Edward Killian, "John Siney: The Pioneer in American Industrial Unionism and Industrial Government" (Ph.d. thesis, University of Wisconsin, 1942).

[26] Bannan's views are cited at length in Gudelunas, "Rise of the Irish Factor."

[27] Gowen charged, in testimony before a Pennsylvania Senate committee at Harrisburg, 8 March 1871, that the Molly Maguires were the enforcers for the W. B. A. See Senate of Pennsylvania, *Report of the Committee . . . in Relation to the Anthracite Coal Difficulties* (Harrisburg, 1871), 13–20. Siney denied it, *Report of the Committee,* 32–35.

[28] Carey-Gardiner Collection, Historical Society of Pennsylvania, Russell to Carey, 19 July 1868.

[29] Senate of Pennsylvania, *Report of the Committee,* 13–20.

[30] Peter Roberts, *Anthracite Coal Communities* (New York, 1904), 18.

[31] Rowland Berthoff, "The Social Order of the Anthracite Region, 1825–1902," *Pennsylvania Magazine of History and Biography,* 89 (1965).

Chapter Two

THEMES FROM IMMIGRANT FRATERNAL LIFE:
THE EARLY DECADES OF THE HAZLETON BASED HUNGARIAN VERHOVAY SICK BENEFIT ASSOCIATION

Bela Vassady, Jr.

To date, little research has been done to analyze the process by which small, local, ethnic mutual-aid associations developed into large national insurance organizations. This paper will discuss selected aspects of that process through the early history of the Verhovay Sick Benefit Association, one of the most successful of the Hungarian immigrant fraternal societies founded during the so-called "new immigration" period (1880–1920). Emerging from modest beginnings in the anthracite region of Pennsylvania during the mid-1880s, the Verhovay Association branched out to become the largest national organization of its kind by 1914, and into a successful life insurance association by 1924.[1] The eight themes associated with the Verhovay experience herein examined are not meant to be exhaustive and leave many important immigrant fraternal roles and functions unexplored. Nonetheless, the themes selected point to important patterns in immigrant institutional history, especially within the much neglected Hungarian-American experience. Some of the themes discussed—such as the mixture of interethnic conflict and cooperation which emerged during the Association's early history; or the relationship which existed between the Association's early anthracite-based fraternalism and its later preference for an anthracite-rooted national leadership; and the ostensibly anti-Semitic tendencies which evolved within the organization from the founders' nationalistically motivated choice of the 'Verhovay' name—point to the development of historical characteristics which were more or less peculiar to the Verhovay experience. Other themes explored—such as the Association's rejection of its intelligentsia leadership, opening the way to the rise of a peasant-rooted leadership with a style of its own; or the consequences of the Association's transient/sojourner membership; and the Association's introduction of democratic institutions which produced much personal and factional conflict in the early decades but led to efficiency and modernization by the 1920s—followed patterns more frequently observed in fraternal societies of other ethnic groups.

Concomitantly, through its examination of selected aspects of the Verhovay experience (1886–1924), this paper also will provide an outline of the developmental stages through which ethnic mutual-aid associations passed as they matured into viable business organizations.

1

A recurring theme in the early years of Hungarian immigrant mutual-aid society formation was the mixture of conflict and cooperation which existed among the multi-ethnic immigrants from northeastern Hungary who pioneered in the anthracite region beginning in the 1880s.[2] This phenomenon was clearly evident in the development of the two earliest Hungarian societies in the region—namely, in the Freeland (Pa.)-based First Hungarian Sick Benefit Society of Pennsylvania, founded in 1884; and, upon the Freeland society's financial collapse and dissolution one year later, in its successor organization, the Hazleton (Pa.)-based Verhovay Sick Benefit Association, founded in 1886.[3]

Established primarily for the purpose of providing mutual aid in times of sickness and death, strong nationalistic motivations also influenced the Hungarian founders of these two early societies. In the anthracite region, Slovaks and Hungarians (Magyars) from multi-ethnic Hungary were indiscriminately lumped together as "Hungarians" by the American public.[4] This confusion proved extremely irritating to affected immigrant groups as they became increasingly aware of their distinct identity and competed for status and recognition in America. Although their common northeastern Hungarian roots and their shared need for security initially led them to join each other's ethnic organizations and to develop cooperative ventures, growing ethnic consciousness soon led to animosity and confrontation between them.[5]

In 1884, the Freeland-based First Hungarian Society expressed its concern over the "bad name" that "barbaric" Slovak behavior was giving to the "real" Hungarians (i.e., the Magyars) in the region.[6] Clarification of the distinction between the two groups was an important goal of the society. This exclusive attitude on the part of the first society at a time when Slovaks far outnumbered the Hungarians may partly explain its swift demise. In contrast, the decision of its successor, the Verhovay Association, to welcome "loyal" (pro-Hungarian) Slovaks to membership provided more recruits, but soon threatened the Association with loss of its Hungarian character. In 1887, the publication of the Verhovay by-laws in Slovak as well as in Hungarian led to a dispute over which language was the official one.⁷ One year later, the earliest minute books of the first Verhovay chapter (founded in Freeland in 1888) were kept in Slovak until the chapter was reorganized along Hungarian lines.⁸ It was a mark of the affinity the two ethnic groups held for each other, as well as their mutual need for survival, that when a violent debate ensued in 1888 over the question of maintaining ethnic and linguistic purity the majority voted to continue Verhovay's multi-ethnic character.⁹ Thereafter, the Association considered itself the "protector" of its "loyal" and "patriotic" Slovak countrymen from the "evil influence" of the "Panslavs," and it assumed a leading role in the struggle against growing Slovak nationalism in the anthracite region.¹⁰ Later, as new Verhovay chapters multiplied in multi-ethnic communities located elsewhere, conflicts with Slovak nationalists increased and the ranks of "loyal" Slovak members were augmented. For example, Chapter 12 in Pittsburgh (founded in 1901) was described as composed of members "mostly with non-Magyar (i.e., Slovak) names, who in their hearts were all good Magyars."¹¹ Not until the vast influx of fresh Hungarian immigrants after the turn of the century assured the Association's continued prosperity did the membership decide that it was no longer necessary to print its by-laws in any language other than Hungarian.¹²

2

A theme prominently reflected in the Verhovay experience was the conflict between the "schooled gentlemen" (the very small minority of educated immigrants) and the "unschooled workingmen" (the peasant-rooted immigrant majority) who insisted upon retaining control over their mutual-aid organizations. Géza Hoffmann has noted that while the immigrants' suspicion of their intelligentsia (pastors, journalists, businessmen) was partially a product of their transplanted resentments against their Old World ruling classes, it more often resulted from their bitter New World experiences with the "business" motivations of their educated compatriots. Thus, while the intelligentsia occasionally participated in initiating early voluntary associations, they (the intelligentsia) were soon removed from leadership positions by the suspicious blue collar majority.¹³ The swift failure of the 1884 Freeland-based society and the immediate founding of its successor, the Verhovay Association, may have been the product of such a move from an "educated" managed society to a "workingman's" society.¹⁴ Thereafter, the desire to retain laborer control and to exclude educated men from leadership became a pattern in Hungarian-American fraternal life. Only one year after its founding, the Verhovay Association's secretary was removed because he was an educated man. His replacement had great difficulty writing the minutes but was more acceptable to the membership.[15] The 1892 expulsion of the Verhovay Association's intellectual advisor, Dr. Árkád Mogyoróssy, after the latter attempted to exploit his close relationship with the founders by urging them to exclude all newspapers from their meetings except his own Hazleton-based publication, *Önállás,* appeared to fall into a similar pattern.[16]

Nor was this pattern modified when the Association achieved national proportions after the turn of the century. When, in 1909, the democratically controlled Verhovay national convention passed a by-law excluding schooled men from serving as officers, it was emulating similar rulings made earlier by other large Hungarian voluntary associations. The Verhovay provision, which excluded pastors, journalists, and businessmen who had "business associations" with the organization, clearly implied fear of exploitation.[17] Despite much opposition from the intelligentsia, the new Verhovay by-law was not repealed until 1911. In the post-World War I years, Verhovay members proved more willing to accept intelligentsia contributions, but the traditional pattern of drawing leadership from working-class roots continued largely unabated into the 1920s.[18] While this practice undoubtedly deprived the early immigrant organizations of experienced leadership, in the long run it provided the mass of peasant-rooted immigrants with the opportunity to develop a self-confident, responsible leadership of their own.

3

Observers of immigrant community life have long noted the intensification of national consciousness among the immigrants after their arrival in the New World. That the pioneer immigrants who founded the Verhovay Association in 1886 were likewise influenced by homeland loyalties and politics was therefore not surprising. However, in naming their Association after a contemporary member of the Hungarian parliament (instead of after a prominent Hungarian historical figure, as did the founders of many other Hungarian-American organizations), they acted in an unusual manner. Their choice went to Julius Verhovay, a parlimentary participant in the short-lived wave of anti-Semitism stimulated by the prominent role played by Jews in the expansion of capitalism in semi-feudal Hungary during the 1880s. Verhovay's name was suggested by Dr. Árkád Mogyoróssy, an immigrant scholar who achieved influence over the founders of the Association by means of his friendship with Michael Pálinkás, the organization's founding father. Retaining a keen interest in the politics of his homeland. Mogyoróssy had maintained close communication with Julius Verhovay and the even more influential anti-Semitic M.P., Goyózó Istóczy, whose name was likewise considered by the founders for their organization. Mogyoróssy apparently convinced the recently arrived northeastern Hungarians that Verhovay best represented their interests back home because he (Verhovay) was "fighting for the welfare of the poor people in Hungary against the big Jewish capitalists and the landowning aristocrats."[19]

In 1892, the Verhovay Association formally voted to exclude Jews from its membership. Repeated proposals of various Verhovay chapters to reverse this exclusion during the coming decades proved unsuccessful.[20] Although the swift growth of the association in the first two decades of the twentieth century demonstrated that this Jewish exclusion did not harm recruitment efforts, it may have had a negative influence when some Hungarian-American Jews emerged as prosperous employers of immigrant labor.[21] Moreover, since the Hungarian-American community generally accepted assimilated (Magyarized) Jews as "good Magyars" and did not usually exclude them from its voluntary association,[22] criticism of alleged Verhovay anti-Semitism (coming especially from the Hungarian-American political left) proved troublesome. Since the Verhovay Association had from its beginnings remained a secular, workingman's fraternal organization with no ideological or political underpinnings, it dropped its anomalous exclusiveness sometime between the mid-1920s and mid-1930s.

4

Although mutual aid remained the primary aim for establishing most fraternal self-help associations, the social and psychological security these organizations provided were equally attractive to potential members. Within these organizations the immigrants found a meaning for existence in their alien environment—a place where they could achieve self-confidence, political and social status, and cultural self-development. Perhaps nowhere was there a greater need for such provisions than in the isolated, scattered anthracite-mining settlements of northeastern Pennsylvania where the Verhovay pioneers labored. The early minute books of the Verhovay Association made repeated references to a strong sense of fraternal belonging reflected in the bonds of brotherhood which developed between the Hungarian miners who joined the organization during its early anthracite years. Instances of members' brotherly support for each other in times of hardship abounded in these early reports, as did examples of the incessant personality squabbles which naturally developed within the intimacy of this closed society of Hungarian miners.[23] Nor were these early bonds of comradship short-lived; as will be described below, these early relationships were to influence the character of leadership within the Association long after it expanded beyond the boundaries of the Pennsylvania anthracite region.

An important symbol of this fraternalism—that of communal drinking together in saloons during meetings—has been one of the most misunderstood and criticized aspects of Hungarian-American behavior. During the anthracite years of the Verhovay Association (1886 to the late 1890s), its regular meetings were held in the Hazleton saloon of John Bugely, one of the founding members. That those who withdrew from sharing the brotherly glass during these meetings were considered unworthy of respect was illustrated by the forced resignation of the president of the Association in 1889 because he refused to drink with the others.[24] After the Association became national in scope, large halls with restaurant and bar services were rented for the purpose of convening the national conventions in various cities. Géza Hoffman, a keen observer of Hungarian immigrant community life, noted that fraternal meetings were held in saloons to provide a congenial atmosphere for socializing and working together. Similarly, the chronicler of the Verhovay Association argued that this practice did not represent a "worship" of drink, but was symbolic of the strong sense of brotherhood and fraternalism which existed in the association.[25] The frequent calls by individual Verhovay members and the Hungarian-American press for an end to heavy drinking during meetings were ignored by the majority of the membership. Although the emphasis on social life, and not on drinking, remained the main purpose for this practice, its perpetuation for decades produced unnecessarily fierce exchanges at conventions which occasionally led to brawls and even police intervention.[26] This all too public Hungarian immigrant fraternal practice was strongly criticized by American observers and was, as a result, frequently lamented by Hungarian-American leaders sensitive to the host society's perception of "shameful" Hungarian immigrant behavior.

5

An important theme in Verhovay fraternal life until after the First World War was the impact that the transient/sojourner character of its membership had on the viability of the organization. The perpetual turnover of membership produced financial instability and almost insurmountable management problems. During the first anthracite decade of the organization, the constant itinerant coming and going of members of the Hazleton "mother society" (founded in 1886) and of its first four chapters (Freeland, Chapter 1, 1881; Mt. Carmel, Chapter 2, 1889; Sheppton and Weston, Chapters 3 and 4, by 1895) reflected this pattern. During this period, sick benefits were covered in a haphazard, ad hoc manner by small fluctuating monthly dues, and death benefits by assessments of the membership at times of death. By 1890, two patterns developed which were to plague the society for decades. These were, first, the "something for nothing" pattern, whereby benefits were continuously raised by majority votes of the membership without commensurate upward adjustments in dues; and second, abuse of benefits, whereby benefits were collected under false pretenses, often with the collusion of the chapter "brotherhood" involved.[27] Such irresponsible behavior is best explained by the itinerant, sojourner character of the members who cared little for the future survival of the organization.[28] Angry debates over these issues notwithstanding, the fear of losing members to competing fraternal societies prevented the Association from doing much about this practice.[29] At its tenth anniversary in 1896, the Association had grown little and was still indistinguishable from the hundreds of ephemeral local societies which eventually outlived their usefulness and disappeared. Still restricted to the mother society and its four anthracite chapters, it had survived by barely surpassing expenses with income. Demonstrating the rapid rate of membership turnover, of 459 members who had joined during these ten years, 304 had departed, thirteen had died, and only 142 remained.[30]

After the turn of the century, the Verhovay Association expanded beyond the anthracite region. This expansion was made possible by two factors: first, the great transiency of the Verhovay members who in moving westward and eastward in their quest for new opportunities founded new chapters wherever they went;[31] and second, the huge influx of new immigrants after 1900 who sought the greater stability and higher benefits of a national organization which furnished them with the means of transferring membership from one city to another. New chapters especially mushroomed in the coal-and steel-producing regions of western Pennsylvania and Ohio, eventually shifting the center of power of the entire organization in that direction.

The itinerant character of the immigrant membership proved to be both a blessing and a curse for the Association. While members cared little for the organization's future and departed in large numbers, new chapters formed and new members appeared even more rapidly than they were lost. As one member of Pittsburgh's Chapter 34 (founded in 1905) reported: "The Hungarians came and went so that during many years officers of some chapters were elected as many as three times during the same year (because they returned that many times) and yet by year's end were again no longer in office (because they again departed)."[32]

The Association grew rapidly after 1903. At its first national convention in 1903, it recorded approximately one thousand members with twenty chapters.[33] Doubled in size by 1906, it was still only one-fourth the size of the leading Hungarian-American fraternal federation.[34] Thereafter, its rate of growth increased phenomenally. By 1914, with 18,203 members and 267 chapters, it far surpassed all other Hungarian-American associations or federations of any size or importance.[35] Although the cessation of immigration during the war years caused a drop in small, local mutual-aid society development, mergers and consolidations produced a thirty-three per cent increase in Verhovay membership from 1914 to 1917.[36] After peaking at 25,169 members in 1920, a period of reform and reorganization finally brought temporary retrenchment by the mid-1920s.

Throughout this period, the Association's Pennsylvania and mid-western orientations—especially in the mining and steel-producing regions of western Pennsylvania and Ohio—were retained.[37] Nor was the predominance of miners within the membership altered. This was reflected in the death and sick benefits paid out during the 1914 to 1917 period: thirty per cent of death benefits were lung-disease related and fifteen per cent were mining-accident related; similarly, forty-five per cent of the sick benefits paid out were for mining accidents.[38]

6

That fraternal societies functioned as "immigrant schools" within which members learned to develop and operate democratic institutions is another important theme clearly evident in the Verhovay experience. The Verhovay "mother society" at Hazleton and its early anthracite chapters held regular monthly meetings at which all of the members debated issues and made decisions by democratic vote. As the Verhovay Association outgrew its anthracite origins, the rank and file displayed great resistance to relinquishing the democratic institutions they had learned to expect in their local fraternals. Thus, a representative form of government based upon the parliamentary convention system was adopted.[39] Delegates sent by the chapters elected national leaders and provided important links between the scattered immigrant communities. Inevitably, however, the system also produced conflicts and tensions between the forces of centralization and decentralization, as the local chapters continued to demand autonomy vis-a-vis their elected national leaders. Thus, while the enforcement of discipline over the behavior of individuals and chapters served as a form of social control within the ethnic association,[40] it also led to interminable hearings and trials in which all actions, including those of the officers were questioned. On all issues, no matter how personal or minor, the delegates insisted that "the only way to peace" was the "democratic way," thereby consuming hours and even days parading witnesses and listening to slanderous counter-charges.[41] Also time-consuming and counter-productive were the long and tedious investigations of the financial dealings of management. Since making allegations against the officers was the quickest way to achieve popularity, these investigations often consumed days.[42]

The political activities within the Association resembled the operation of the parliamentary system in a national government. For months before conventions met, agitation and political campaigning were evident at the grassroots level as various parties developed. The ethnic press fanned the flames of agitation as it sided with various leaders and factions and printed polemical personal exchanges. Because they set the tone for forthcoming events, welcoming banquets held the night before conventions were closely watched affairs. Here, as one observer noted, the "old gladiators" faced one another "prepared to do battle."[43] The highly politicized delegates were sent to the conventions with strict instructions to avoid costly reforms and to support certain factions, and their continued popularity at home depended upon their ability to be heard and noticed in the ethnic press. After days of constitutional debates punctuated by loud and flamboyant speeches, decisions were made by majority vote of the delegates. Since by 1920 Verhovay decisions impacted upon over three hundred Hungarian-American settlements in the United States (there were that many chapters by then), they were followed with great interest by the entire immigrant community.

7

Closely linked with the enthusiastic practice of democracy was the self-confident "leadership style" which developed within this milieu. Verhovay leadership was derived from among mobile men of peasant stock. Although without formal education, many had migrated to America with the skills of tradesmen or artisans. Those who emerged as officers in the national organization were self-made individuals, men who had gained status and respect by moving from humble beginnings as physical laborers into private businesses they had initiated to serve the ethnic community. Although forming a new middle class, they retained their legitimacy among the peasant-rooted immigrants by virtue of their shared working-class origins.[44] Gaining experience as organizers and officers of local fraternal societies, they moved to the national level, where they provided a loose and fragmented form of self-government for the wider ethnic community. They learned to stir the hearts of fraternal memberships with a bombastic, flourishing oratorical style which appealed to audiences of humble origins proud of their recent achievements in literacy and status.[45] Similarly, they exploited a symbiotic relationship existing between the fraternal societies and the ethnic press. The newspapers competed for the opportunity to serve as "official newspapers" for large societies. Selection to this status guaranteed that the fraternal membership would subscribe to the chosen "official" paper. The "official newspaper" became an uncritical mouthpiece for the national leadership of the society and exercised inordinate influence upon the organization's internal politics.[46] Whether on the convention floor, at the banquet table, or in the ethnic press, these "unscrupulous demagogues," as Kende referred to the leadership, often took on dictatorial airs as they exploited two popular themes which guaranteed to win them mass support—the traditional fear and suspicion of the "educated gentlemen," and the fashionable "something for nothing" campaign promise.[47]

This leadership style was perhaps best exemplified by Michael Pálinkás, the Hazleton miner who as founder and long-time first president of the Association controlled it for over a decade.[48] Exhibiting the typical transiency of Hungarian immigrant miners during the 1880s and 1890s, Pálinkás repeatedly resigned his office to try his luck elsewhere only to be reinstated by popular demand each time he returned to Hazleton. When he finally moved permanently to the west at the turn of the century, he served as the Association's first national secretary and founded no less than five new Verhovay chapters in Ohio.[49] Of these, Chapter 14 in Cleveland was destined to become one of the largest and politically most influential of the Verhovay lodges. Like many of his contemporaries, Pálinkás repeatedly found himself implicated in allegations of misconduct and mismanagement, resulting in his suspension on several occasions.[50] But his popularity with the brotherhood prevailed. He returned again and again to national conventions as a delegate representing various chapters he had founded,[51] and continued to be elected to significant offices during the first two decades of the twentieth century.[52]

Until 1909, top leadership remained in the hands of a coterie of men closely associated with the Hazleton mother society and its anthracite chapters. The 1904 national convention chose Joseph Arnoczky (Freeland, Chapter 1) as president and Andrew Buczko (Mt. Carmel, Chapter 2) as treasurer. Arnoczoky was re-elected in 1905, while Vladimir Deák (Hazleton, mother society) was elected as secretary and Andrew Bolla (McAdoo, Chapter 11) as treasurer.[53] Two years later the same group of anthracite men was returned to office, but the selection of Alexander Kövér (Johnstown, Chapter 8) and Béla Perényi (New York, Chapter 38) to the influential auditing committee, which investigated management's financial performance, pointed to the rise of new men who were soon to challenge the old guard.

The anthracite clique was successfully removed in 1909, a year of crisis marked by demagoguery, mismanagement, and factionalism. Three conventions (the regular one in June and two extraordinary ones in November and December) were required in 1909 to save the organization from dissolution. As chairman of the investigative auditing committee at the June convention, Perényi led a reform faction which accused Bolla, the treasurer, of mishandling fraternal assets. Although four days of bitter debate exonerated Bolla on the grounds that he had "not intentionally" done wrong, the incident led to lingering suspicions.[54] With obvious political ambitions of his own, Perényi also engaged in violent floor battles with Emery Fecsó, the Cleveland-based publisher of the Association's "official newspaper," the *Magyar Hírmondó*.[55] Fecsó, who may have had strong anthracite connections,[56] was accused of conspiring with Arnoczky and his supporters to re-elect the *Hírmondó* as "official newspaper." That Arnoczky managed to do just that, despite a mood of opposition among the delegates, only added to the anger and suspicion about his authoritarian methods which led to his losing the presidency at the end of the meeting.[57] But Deák retained his office, as did Bolla, notwithstanding his obvious record of mismanagement. The presidency went to Kövér, who, after joining the Hazleton mother society in 1896 (thereby also establishing an anthracite connection), had followed the common pattern of moving west and helping to start the chapter he now represented.[58] Kövér was to preside over the Association for nearly a decade.

The delegates of the June 1909 convention returned to their constituents amidst lingering suspicions about mismanagement and authoritarian actions. Agitation within the chapters reached a high pitch, with some chapters publicly rejecting the "official newspaper" and others quitting the Association altogether.[59] Printing rumors and personal attacks, the ethnic newspapers took up the battle as they competed for the expected change in "official newspaper" status.[60] Claiming to be heading-off an imminent threat of dissolution, Perényi called for an extraordinary convention to be held under the auspices of the two New York chapters (38 and 83) and received the strong support of the disgruntled anthracite chapters. Reflecting their customary opposition to central authority, approximately one-half of the chapter delegates who attended the November convention in New York denounced the presiding officers (especially Kövér, Deák had resigned) for disciplining various individuals and chapters in an "unconstitutional and authoritarian manner" since the June convention. New elections placed three anthracite men, including Arnoczky and Deák back in office. Only the treasury went to a non-anthracite man.[61]

The extraordinary convention held in New York resolved nothing. At a third convention called by Kövér in December, the actions of the presiding officers were vindicated and the "official newspaper" issue was resolved by the selection of more than one newspaper for this purpose.[62] After conducting a series of "trials," the alleged "conspirators"—Perényi, Deák, and others implicated—were expelled from the fraternal organization. Arnoczky was refused recognition as a duly constituted delegate and was ejected; Bolla resigned.[63] The post of treasurer and secretary were filled by Bertalan Ranky (Homestead, Chapter 89) and Stephen Gábor of Lorain, Ohio, respectively.[64] The new officers, including Kövér, reflected a shifting of influence to the growing number of new chapters in western Pennsylvania and Ohio, which was further confirmed by a long and bitter battle over transferring the home office from Hazleton to Pittsburgh (an issue not to be finally resolved in Pittsburgh's favor until 1926).[65]

8

The persistant resistance to reform by the "traditionalists" (i.e., the transient/sojourner membership), until the victory of the "modernizers" in 1923, will be the final Verhovay theme examined in this study. Although more experienced in financial management than their predecessors, the new officers corps elected in 1909 continued to ignore the changes necessary to convert the rapidly growing society from its traditional fraternal emphasis to a viable business concern. Given the continued sojourner mentality of the membership, to do otherwise would have been suicidal for the officers.[66] For this reason, benefits after 1909 continued to be raised without commensurate upward adjustments in dues; dues continued to be uniformly applied regardless of age at time of application; and control over new memberships and the system of benefit payments remained very lax.[67] The lone reform advocate was the secretary, Stephen Gábor, whose proposals in favor of re-rating all membership dues based upon chronological age were ignored for a full decade after he first made them in 1914.[68]

Although by 1917, at the most chaotic and hotly disputed convention since 1909, the pressure to reform had intensified, it was politics as usual as the traditional campaigning for new leadership resumed. The New Yorker, Gyula Völgyi (Chapter 38), whose political influence had been rising during the previous two conventions of 1911 and 1914,[69] was elected to chair the 1917 convention by a well-organized, vociferous majority which then engineered the defeat of the "old guard" (Kövér's party) and swept Völgyi into the presidency.[70] A third faction, composed of a small minority of modernizers led by Joseph Daragó, who supported Gábor's urgent calls for chronological age premiums and other necessary fiscal reforms, found itself ignored by the anti-reform majority.[71] Although a grudgingly conceded "compromise" agreement imposing a modest across-the-board rise in dues was finally accepted by a majority of the membership, it did little to respond to the crisis proportions reached by the Association's growing indebtedness.[72] The adoption of an in-house news organ, the *Verhovayak Lapja,* to replace the old "official" newspaper system, emerged as the only notable reform decision made at the 1917 convention.

The years from 1917 to 1923 saw the battle between the "traditionalists" and the "modernizers" finally culminated in victory for the latter. While in 1917 the majority of Verhovay members had still reflected the sojourner mentality of impermanence, at the close of the war the realities of changed political and social conditions were beginning to dawn upon many. As they came to accept their stay as permanent, they also recognized that the modernization of the Association was now to their advantage.[73] Pressures to modernize came from inside and outside the organization. The slower turnover of membership due to increasing permanence naturally produced older average age levels and higher sickness and death rates. As a result, the life insurance component of the Association's services, which until now had been secondary to its sick-benefit function and had been funded by assessed contributions and inadequate dues, experienced a financial crisis.[74] External pressures to modernize were also compelling. One state government after the other withdrew operating privileges from Verhovay chapters until they complied with new state insurance standards.[75] For the Verhovay Association the mandate was clear—reform or die.

But the necessary reform did not come easily. The degree to which the leadership, now converted to the need for modernization, remained at the mercy of the anti-reform party was demonstrated in 1917 and again at the 1920 convention. The 1920 delegates came with the usual orders from their chapters to resist change. After listening with impatience and boredom to long reports describing deficits and reform needs, the delegates sprang to life when the time arrived for allegations against the officers. But in response to this traditional form of obstructionism, the officers this time took the untraditional step of going on a two-day strike. After other attempts at obstructionism, the delegates finally accepted the chronological age-scale concept "in principle," but the scale they adopted had no mathematical basis and still protected the older members most hurt by the re-rating process.[76]

But the days of half-way measures were over. Dissolution was threatened as more states withdrew operating privileges from Verhovay chapters. By 1921 the pro-modernizing president, Joseph Daragó, was forced to resign over the unresolved reform question, and was followed in the coming year by a continuous turn-over of officers in face of the stubborn resistance to reform.[77] The necessary reforms finally came at the 1923 convention. In addition to streamlining the entire convention system to provide for future managerial efficiency,[78] the Association adopted a scale of premiums based upon mathematically calculated actuarial mortality tables.[79] Although this re-rating process caused the temporary departure of approximately seven thousand older Verhovay members, it assured the survival of the Association as the viable insurance company it has remained to the present day.[80]

NOTES

[1]Source material for researching this period can be found in the Pittsburgh offices of the William Penn Fraternal Aid Association, the corporate name under which the Verhovay Association presently operates. Although the Verhovay minute books were destroyed in 1936, two alternative sources provide satisfactory substitutes: the fiftieth anniversary history of the organization, Joseph Daragó (ed.), *Verhovayak Lapia: Ver-*

hovay Segélyegylet 50 éves jubileumi kiadvanya (Pittsburgh, 1936), hereafter cited as *V. L.,* and the contemporary accounts in the ethnic press, especially in the New York daily, *Amerikai Magyarok Népszava,* hereafter cited as *A. M. N.* Beyond that, especially useful are the insights of two contemporary observers: Géza Kende, *Magyarok Amerikában,* 2 vols. (Cleveland, 1927); and Géza Hoffmann, *Csonka munkásosztály. Az amerikai magyarság* (Budapest, 1911).

[2]Early reports refer to Hazleton as the "mecca" of immigration from northeastern Hungary in the 1880s—especially from Sáros County, but also from neighboring Zemplén, Ung, and Abauj Counties. Hazleton served as a dispersal center for the transient immigrants whose numbers fluctuated wildly. See Kende, *Magyarok,* 2:43; Tihamér Koháffnyi, *Az amerikai magyarság múltja, jelene és jövője. A "Szabadság" tizéves jubileumára* (Cleveland, 25 December 1901), 11. Hereafter cited as: *Szabadság 10th Anniversary.*

[3]There is strong evidence of close association between the earlier Freeland society and its Hazleton-based successor. A founding member of both societies stated that the Hazletonians, who comprised the majority of the first society, insisted on moving their headquarters from Freeland to Hazleton. (See Joseph Uhlár to Stephen Gábor, 2 February 1911, Freeland, quoted in *V. L.,* 43). Of those named on the list of twenty-eight original Verhovay subscribers and thirteen charter signers in 1886, several had participated in the founding of the Freeland society, but with the exception of one (Uhlár), all were Hazletonians. Further evidence of the close association between the first society and its successor is provided by the following facts: the treasury of the first society was transferred to its successor, some of the same individuals served as officers in both organizations, and the first Verhovay chapter was founded in Freeland in 1888. *V. L.* 25–26; *Szabadság,* 21 December 1911, part 5, 2.

[4]To illustrate, in the otherwise excellent survey by Peter Roberts, *Anthracite Coal Communities* (New York, 1904), the author confuses ethnic terms such as Magyar, Hun, Hungarian, and Slav (pp. 21, 24, 28, 32). Similarly, in a more recent work by Victor Greene, *The Slavic Community on Strike* (Notre Dame, 1968), the author excludes Magyars from the ranks of eastern European miners in the anthracite region, inaccurately referring to all of them as Slavs. Greene identified John Németh, a prominent Hungarian ship ticket agent, banker, and consul agent in Hazleton as a Pole (pp. 48, 142). In fact, Németh had come to Hazleton as a miner in 1879 at the age of eighteen from Abauj County in Hungary. He was a founding member of the Freeland- and Hazleton-based Hungarian associations and served as an officer in both.

[5]Since Slovaks greatly outnumbered the Magyars immigrating from Hungary in the early years, most of the early societies and churches in Hazleton and other anthracite communities were Slovak-initiated and controlled. But initially the Slovaks identified with their Hungarian homeland and many of them were bilingual, thus attracting Hungarian participants with whom they often developed cooperative ventures. The mining companies and Roman Catholic bishops also perceived all of these immigrants as "Hungarians" and insisted that they work together. Kende, *Magyarok,* 2:80–82; *Szabadság,* 21 December 1911, part 5, 2.

[6]*V. L.,* 58–59. The minute book specifically complained about a "barbaric ball" held by the Slovaks in December 1883 under the Hungarian name. In 1901, Tihamér Koháffnyi, editor of *Szabadság,* complained that when Slovaks were arrested in Hazleton they identified themselves as Hungarians, thereby giving Hungarians a bad name. *Szabadság, 10th Anniversary,* 11.

[7]The Hungarian version claimed that Hungarian was the official language, while the Slovak version made the same claim for itself. *Szabadság,* 21 December 1911, part 5, 2.

[8]*Ibid.*
[9]*V. L.,* 44, 60.
[10]In 1889, the mixed Hungarian and Slovak membership of Verhovay and the neighboring St. László Workers Sick Benefit Society of Audenried made an unsuccessful attempt to merge for the purpose of "putting an end to our enemies, the Panslavs." *Amerikai Nemzetör,* 8 December 1889, quoted in Kende, *Magyarok* 2:312. An observer in the 1890s described Verhovay as the major anti-Panslav organization in the anthracite region. Loránt Hegedüs, *A magyarok kivándorlása Amerikába* (Budapest, 1899), 35.
[11]*V. L.,* appendix, 11.
[12]*Szabadság,* 21 December 1911, part 5, 2.
[13]Hoffmann, *Csonka,* 177–179; Kende, *Magyarok,* 2:273–274.
[14]*V. L.,* 24, 43, 64; Kende, *Magyarok,* 2:315–316.
[15]*V. L.,* 44.
[16]*Ibid.,* 45.
[17]*A.M.N.,* 14 June 1909. The precedent was set in 1897 by the largest (at the time) of the Hungarian societies, the Sick Benefit Societies Federation of Bridgeport, which restricted leadership to "working men," whom it defined as men known to have engaged in physical labor in the Old World. (Hoffmann, *Csonka,* 177–178; Kende, *Magyarok,* 2:316).
[18]In the first decade of the twentieth century, clergymen were occasionally selected as chapter delegates. Although only in a temporary role, the earliest important post went to the Reverend Alexander Kalassay of Pittsburgh when he was chosen to chair the 1914 convention. The election of the young lawyer, Géza Bruger, to the post of treasurer in 1917 was the first break-through in utilizing an educated professional in a major elected office. Since of all the intelligentsia, journalists were most suspected of exploiting the immigrant community, the selection of the newspaper editor Alexander Zámbory to chair the 1920 convention was truly exceptional; it moved another journalist to exclaim in surprise: "It is not a crime after all to be a trousered gentleman (nadrágos ember) in the Verhovay Association." *A.M.N.* 19 May 1920.
[19]Júlianna Puskás, *Kivándorló magyarok az Egyesült Allamokban, 1880–1940* (Budapest, 1982), 224–225. That Mogyoróssy was politically and not racially motivated is suggested by Edmund Vasváry's comments about this incident in the Vasváry Collection. Further evidence for this interpretation of Mogyoróssy's motive is reflected in Mogyoróssy's letter of 26 December 1893 to Pálinkás, in which he recommended that Jews be permitted to organize a chapter of their own in a new Pennsylvania Hungarian fraternal association he and Pálinkás were contemplating starting. See the Edmund Vasváry Collection, Hungarian Studies Foundation, New Brunswick, New Jersey, Micro. Roll #7, which includes a biography of Mogyoróssy by Pálinkás, letters exchanged by Mogyoróssy and Pálinkás, and Vasvary's comments on the whole affair. Also see *V. L.,* 25.
[20]*V. L.,* 28; *A.M.N.* 20 June 1911, 31 Oct. 1917, 28 May 1920.
[21]*V. L.,* appendix, p. XI.
[22]Yeskayahn Jelinek, "Self-Identification of First Generation Hungarian Jewish Immigrants," *American Jewish Historical Quarterly,* 61 (March 1972), 214–222.
[23]*V. L.,* 65.
[24]*V. L.,* 44–45, 65.
[25]Hoffmann, *Csonka,* 142; *V. L.,* 66.
[26]For examples of heavy drinking and violence at meetings, including police intervention in at least two instances, see: *V. L.,* 44–46; *AMN,* 26 June 1909, 22 July 1914, 19 May 1920.

[27] *V. L.,* 27.
[28] Kende, *Magyarok,* 2:277.
[29] *V. L.,* 28; Hoffmann, *Csonka,* 169
[30] Kende, *Magyarok,* 2:312.
[31] The first major movement from the anthracite region was to Ohio, where one member founded five new chapters at the turn of the century. New chapters were also founded by itinerant anthracite Verhovay men in western Pennsylvania and New Jersey. (*V. L.,* appendix, pp. I & II.) A minimum of eighteen Hungarian immigrants were required to start a new chapter.
[32] *V. L.,* appendix, p. V.
[33] *Szabadság,* 21 December 1911, part 5, 2. One year later it was up by 20% to 1200 members and 25 chapters. (*Szabadság Naptára,* 1905, 207).
[34] In 1906 it claimed two thousand members. This was one-fourth the size of the Sick Benefit Societies Federation of Bridgeport (*Szabadság Naptara,* 1907, 103). For further comparative statistics, see Hoffmann, *Csonka,* 158–161.
[35] The statistics of growth are as follows: 1907—5,116 members and fifty-three chapters; 1909—7,200 members and 145 chapters; 1911—9,333 members and 174 chapters; 1914—18,203 members and 267 chapters.
[36] For 1917, the Association reported 24,337 members and 298 chapters.
[37] The breakdown of 264 chapters in 1915 was as follows: 41.6% (110) Pa.; 20% (53) Ohio; (17) Ill.; (16) N.J.; (16) W. Va.; (12) Ind., (9) N.Y.: (8) Mich.; (5) Conn. (See: *Verhovay Annual Statement* (Hazleton, 1915). In 1922, the breakdown of the 21,845 membership was as follows: approx. one-third (7,425) Pa.; approx one-fourth (5,725) Ohio; (1,826) N.J. (See: *Verhovayak Lapja,* 30 June 1922.)
[38] *A. M. N.,* 23 October 1917.
[39] Contemporary press reports offer detailed descriptions of how this system worked. Also see: Kende, *Magyorak* 2:261–267, 315–316; and Hoffmann, *Csonka,* 175–176.
[40] As early as 1888, Verhovay members who made violent threats upon fellow members were expelled from the organization (*V. L.,* 44). The fraternal rules refused benefits for sickness or death resulting from brawling, drunkenness, or venereal disease. Hoffmann, *Csonka,* 153–54, 167, 175–176.
[41] Quoted in *A. M. N.,* 27 July 1914. For other examples of such hearings and trials, see: *A. M. N.,* 12 June 1909; 1 & 2 December 1909; 20 June 1911; 20, 22 July 1914; 27, 29 October 1917.
[42] For example, see: *A. M. N.,* 16 June 1911; 25 & 27 July 1914; 24 & 25 May 1920.
[43] *A. M. N.,* 27 July 1914; Kende, *Magyarok,* 2:506. Kende reminisced about how the order in which the toastmaster called upon speakers at these banquets reflected much about protocol and the nature of the upcoming power struggles. American officials were always invited for the purpose of gaining recognition and status for the Hungarian Verhovay community. (See: *A.M.N.,* 19 May 1920.)
[44] Kende, *Magyarok,* 2:271–273.
[45] *Ibid.,* 2:261–262.
[46] On the role of the press, see Hoffmann, *Csonka,* 184; Kende, *Magyarok,* 2:- 266–271, 307–308; *A.M.N.,* 26 June 1909.
[47] Kende, *Magyarok,* 2:305–306, 315–316.
[48] Pálinkás, who had also been a participant in the administration and disbanding of the earlier Freeland society, was the highest contributor to the founding of the Verhovay Association with two dollars (most of the founders contributed fifty cents). Trained as a skilled cabinetmaker in Hungary, he emigrated from Sopron County to

Hazleton in 1882 at the age of twenty-one. His seventeen years in the Hazleton mines were broken by several unsuccessful attempts to go into private business. From 1886 to 1899, after which he moved to Ohio, he served (with some minor breaks) as Verhovay president. See: *V. L.,* 25; Kende, *Magyarok,* 2:66; Vasváry Collection, Roll #8; Kálmán Káldor (ed.), *Magyar-Amerika irásban es képhen* (St. Louis, 1937), 1:58.

[49] *V. L.,* 27–28; *Ibid.,* appendix, pp. II & III; Kende, *Magyarok,* 2:318.

[50] In 1904, he was accused of stealing the minute books (1889–1904) of the Association to destroy evidence of his own misconduct. He was also accused of mishandling the funds of Chapter 14 in Cleveland, *V. L.,* 28.

[51] *V. L.,* 45–46; *A. M. N.,* 3 June 1907; *Szabadság,* 21 June 1914; *A. M. N.,* 4 October 1923.

[52] In 1914, he missed being elected president by a close vote and then won the vice-presidency by acclamation (*A. M. N.,* 1 August 1914). He continued to be nominated to top posts and held important positions in the coming years. By the mid-1930s he was lionized as the respected founding father of the Association and of the influential Chapter 14 of Cleveland.

[53] *V. L.,* 29; *ibid.,* Appendix, p. II; *A. M. N.,* 17, 19 November 1909. Buczko and Bolla had both served as treasurer in their respective chapters. Deák had been president of the Hazleton mother society.

[54] It was discovered that Bolla had deposited the Association's funds in his own name in a McAdoo bank of which he was an officer. His defense was that this did not appear inappropriate to him. *A. M. N.,* 10 June 1909.

[55] *Ibid.,* 14 June 1909; Kende, *Magyarok,* 2:314–315.

[56] According to Kende, Fecsó was one of the founders of the Freeland society in 1884. By the 1890s Fecsó had moved west and was the owner of a Hungarian saloon in Cleveland which served as a center for immigrant political and cultural activities. At the turn of the century he was the publisher of two newspapers, the *Magyar Hírmondó* and the *Magyar Napilap,* both of Cleveland. Kende, *Magyarok,* 2:126–127.

[57] *A. M. N.,* 14 June 1909.

[58] *V. L.,* 26; *Ibid.,* appendix, p. II.

[59] *A. M. N.,* 31 July, 14 October, 1 November 1909. The chapters complained about discipline from the central office which had resulted in the expulsion of several members. Involved was the old problem of centralization versus the demand for autonomy by the local chapters.

[60] As a result of losing Verhovay support, the *Magyar Hírmondó* went bankrupt in the fall of 1909. The two great dailies, *A. M. N.* and *Szabadság,* competed for the "official newspaper" designation. *A. M. N.,* 11 October 1909.

[61] *A. M. N.,* 16, 17, 19 November 1909.

[62] Kende, *Magyarok,* 2:267–271; 313–318.

[63] *A. M. N.,* 20 November, 1–3 December 1909.

[64] Ranky had served as treasurer of his chapter and was a local banker in Homestead. Gábor emigrated at the age of 25 from Abauj-Torna in 1895 with the skills of a machinist. Starting as a laborer in Cleveland, he founded a voluntary association and served others before he moved to Lorain, Ohio, where he began a small business. He was a primary mover for change in the Verhovay Association during the several decades that he served as its secretary after 1909. See Kaldor, *Magyar-Amerika,* 1:245; Vasváry Collection, Roll #3.

[65] Although by 1909 the westward shift of membership had become evident, the sentimental and political influence of the anthracite chapters remained very strong. Conventions from 1911 through 1923 spent hours debating and voting on the issue of

transferring the home office. Chapter 11 (McAdoo) led an anthracite centered legal suit against the Association after the majority of the delegates finally agreed to transfer the home office to Pittsburgh in 1914. The resulting court proceedings forced the headquarters to be returned to Hazleton in 1915, where it remained until legal questions and membership sentiments were finally resolved by 1926. See *A. M. N.,* 20 June 1911; 29, 30 July, 8 August 1914; *Szabadság Naptára* (1915), 149; *A. M. N.,* 23 October 1917; 28 May 1920; 20 October 1923; *V. L.,* 36.

[66] *V. L.,* 30.

[67] *A. M. N.,* 22 June 1911; 20, 31 July 1914; *V. L.,* 30.

[68] See Gábor's report in *A. M. N.,* 20 July 1914.

[69] At the 1911 and 1914 conventions, the rising political prominence of Gyula Völgyi had become evident. Like others before him, Völgyi had adroitly exploited the chairmanship of the influential auditing committee which investigated management and proposed changes. See, for example, *A. M. N.,* 23 June 1911.

[70] *A. M. N.,* 23 October, 3 November 1917; 18, 19 May 1920; *V. L.,* 33. Less than one year later Völgyi was removed from office for overstepping his powers while pursuing a reinsurance agreement with an American insurance company. The reinsurance agreement, signed later in 1918 by Völgyi's successor, Joseph Daragó, saved the Association from certain bankruptcy after the influenza epidemic took nearly five hundred Verhovay lives during the coming year.

[71] *A. M. N.,* 1 November 1917; Kende, *Magyarok,* 2:319; *V. L.,* 34. The minority was led by Joseph Daragó, an Akron, Ohio butcher-shop owner of peasant background from Hungary. He was elected vice-president in 1917 and, as noted in the previous footnote, succeeded to the presidency in 1918.

[72] *A. M. N.,* 23, 24 October, 1 November 1917; *Szabadság Naptára* (1918), 100–101.

[73] *V. L.,* 67.

[74] See Gábor's report for 1917 in which he explains how conditions had changed and notes that death benefit payments rose 94.7% from 1914 to 1917. *A. M. N.,* 23, 24 October 1917.

[75] *A. M. N.,* 1, 4 October 1923.

[76] *A. M. N.,* 22, 27 May 1920; *V. L.,* 34.

[77] *V. L.,* 34.

[78] By the early 1920s, the inefficiency of over three hundred delegates "democratically" deciding all issues could no longer be tolerated. As the number of chapters had grown, so had the delegates sent to represent them, until the length and cost of the conventions had become prohibitive. Reform attempts, such as increasing the time between conventions, had done little to solve the problem. (For example, the 1914 convention consumed an unprecedented eleven working days at a cost of $20,000, more than one dollar per member. See *A. M. N.,* 31 July and 1, 8 August 1914). Reforms made in 1923 finally terminated most of these problems. The chapters were grouped into twenty-two districts, each represented by two delegates. The conventions thereafter made decisions quickly and professionally, saving money and time, while retaining a form of representative democracy.

[79] *A. M. N.,* 4, 9 October 1923.

[80] Kende, *Magyarok,* 2:321–322; *V. L.,* 35; *A. M. N.,* 10 October 1923; *A. M. N.,* 14 September 1925.

Chapter Three

COMMENTARY: THE MINERS OF ST. CLAIR AND THEMES FROM IMMIGRANT FRATERNAL LIFE

Joseph T. Makarewicz

The rise in ethnic consciousness has aroused a great deal of interest in ethnic intergroup and intragroup relations. The collection of essays found in *The Ethnic Experience in Pennsylvania,* edited by John E. Bodnar, is a classic example of the type of literature generated by this ethnic consciousness. The papers of Professor Anthony F. C. Wallace, "The Miners of St. Clair: Family, Class, and Ethnicity in a Mining Town in Schuylkill County, 1850–1900," and Bela Vassady, Jr., "Themes from Immigrant Fraternal Life: The Early Decades of the Hazleton Based Hungarian Verhovay Sick Benefit Association," describe the conflicts faced by immigrants in their relations with one another and with the dominant local culture.

Professor Wallace divides his paper into three parts. In the first part he provides us with an interesting thesis, viz., that during the period from 1850 to 1880, ethnic groups in the anthracite region shared a common culture and "that heightened consciousness of ethnic differences, and venting of ethnic antagonisms between strikers and strikebreakers, was artificially induced in the course of social conflicts that had their source in economic rather than ethnic competition."

I would like to suggest that this thesis may be extended beyond the anthracite region and possibly to the iron and steel manufacturing regions around Pittsburgh, such as Aliquippa, Ambridge, Braddock, Homestead, Midland, and Monessen. In my studies of ethnic groups and their role in the unionization drives of the 1930s in these areas, especially in Aliquippa, I have found that antagonisms and conflict among the various ethnic groups were artificially induced and were not a natural state of affairs. There are numerous examples of ethnic groups sharing such facilities as churches, halls, and other buildings for their activities. Management of the mills, to prevent cooperation among the ethnic groups in seeking economic and social improvement, attempted with diminishing success in the late 1920s to keep workers divided. It seemed to be a matter of management policy to keep ethnic groups separated in the mill and to threaten to replace those groups and individuals who made trouble with others of a different ethnic group, thus fostering competition for jobs and, at times, hostilities between members of one group and another.

Professor Wallace resists the simplistic analytic scheme that the existence of ethnicity was responsible for a two-class society. He suggests instead "that ethnicity only became a source of cleavage and alliance after economic and class conflicts had become intense." There are four reasons, according to Professor Wallace, for the lack of early ethnic conflict: First, is the ethnic homogeneity of the early coal miners; they were English, Welsh, Scots, and Irish, with the English language as the common bond. Second, is the common culture and technology in mining coal; once the immigrant acquired the requisite knowledge and information, he advanced from miner's helper, with its lower pay and rank, to "miner" itself. Thirdly, these early groups believed in the possibility of advancement because they were familiar with what Professor Wallace refers to as the "map of regional organization." What he means, I think, is that immigrants quickly learned the "ropes" of the economic and political network that managed the whole system. Fourth, the kinship system in ideology and in practice was similar.

In explaining this last point, Professor Wallace indicates that there were a small number of inter-ethnic marriages; however, he doesn't indicate whether these marriages resulted in a change in social status of one or the other of the partners, or both. Were the inter-ethnic marriages the result of a lack of males or females within a particular ethnic group? If not, what forces were at work in bringing them about? It would be interesting to know the number of inter-faith and inter-ethnic marriages. Such information would shed light, perhaps, on the degree of inter-ethnic rivalry and social interaction. Professor Wallace states that the religious conflict among Catholic and Protestant miners was instigated by capitalists in an attempt to divide the workers. Unfortunately there isn't enough information about inter-ethnic relations.

Professor Wallace, however, views the conflict between ethnic groups with a broader perspective than just a struggle between Catholic and Protestant, and worker and capitalist; he regards it as a struggle for survival among small entreprenuers, the question being who should survive and what kind of culture will emerge. He has thus set the stage for an interesting discussion of how class struggle and ethnicity were used by the small entreprenuerial class to stave off collapse in the face of modernization changes in the economy and technology during the mid- to second-half of the nineteenth century. Once again, however, he does not describe these new forces which were bringing about this class struggle.

In the second part of the paper he compares the relationship of ethnicity, occupation, and family. Professor Wallace arrives at some interesting conclusions from his comparison of these three variables in the ethnic population census of 1850 and that of 1870. He concludes that there is a direct relationship to one's ethnic identity and occupation and that there was a significant upward mobility among the Irish working in the mines of St. Clair. That a number of Irish workers went from laborers to contract miners is regarded as a sign of social improvement among the Irish. The statistics to show this are sparse, however; the use of city directories, tax lists, and court records may shed more light on this conclusion. Until more information is available, I think that Professor Wallace's conclusions should be regarded with caution.

The final section of the paper returns to an earlier point, viz., that business failure, mismanagement, and "criminal elements" among the working class were responsible, at least in the eyes of some members of the community, for the social conflict in the anthracite community. Professor Wallace states that the ethnic diversity of the anthracite community was regarded as being responsible for the problems of violence and economic collapse of small collieries, and that the Irish, identified with the Molly Maguires, were the criminal elements in the region responsible for all the problems. Here we have a good example of how immigrants were frequently made the scapegoats for economic and social problems of the day, and how any attempt by them to improve their position was regarded as a conspiracy to disrupt society.

The purpose of fostering this ethnic conflict, according to the author, was to "tilt" middle-class sympathy away from the miners and toward legislative and administrative action in favor of the mine operators. This is an interesting and, if it can be further documented, an important conclusion for the study of ethnic-group relations in America. I don't believe Professor Wallace has given us enough evidence to fully justify this conclusion, but my own studies of inter-ethnic group relations in Aliquippa and Ambridge, Pennsylvania in the 1930s seem to confirm his conclusion. We must be careful, however, not to substitute one conspiracy theory with another.

The value of this study is that it opens some interesting avenues of research for those interested in questions related to inter-ethnic rivalry and its relation to the history of labor and capital in America.

It is acknowledged that the very first fraternal benefit society in America was the Ancient Order of United Workmen, which began in Meadville, Pennsylvania on October 27, 1868. It was organized by John Jordan Upchurch, a railroad master mechanic. Upchurch believed that labor and management were intolerant of each other because neither could see the other's viewpoint. He sought a plan that would obviate strikes and improve labor-management relations. M. W. Sackett, in his history of fraternal benefit societies, quotes a letter by Upchurch in which the latter wrote: "In my mind the best means to accomplish this objective was to bring together employers and employees face to face, by uniting them in the bonds of fraternal friendship."

Included in the Upchurch program was a plan to provide workingmen with protection for their families. He felt that such protection had been denied them because insurance in commercial companies cost more than most workingmen could afford.

The idea of fraternal protection was novel in that day and, because it was simple and inexpensive as well as fraternal, the members of such societies multiplied rapidly.

The development of the fraternal movement in America coincided with the period of great immigration from foreign countries to this land between 1880 and 1920. The fraternals acted as a channel through which new immigrants at the outer border of society entered into the mainstream of American democracy.

Crowded together in the cities, as most of them were, the immigrants felt more comfortable and less alone in the ethnic societies they organized. But, as they took part in their societies' proceedings, they learned that they had freedom to express opinions without fear, and they found that they had the right to dissent and the obligation to vote. They learned the democratic process by observing and participating in it in the meetings of their fraternities.

Forty million people had been assimilated into this country during those forty years, and never before had immigrants so rapidly become an integrated force in a political system.

Fraternal societies like the Hungarian Verhovay Sick Benefit Association deserve much of the credit for that accomplishment. Thousands of immigrants coming together regularly in church meeting rooms and basements and conducting their affairs in a foreign language, learned—sometimes painfully—the privileges and responsibilities of freedom in America.

Mr. Vassady's paper, unfortunately, does not live up to the expectations of this title: "Themes from Immigrant Fraternal Life." He seems to stress the negative sides of the Verhovay Association, such as it's anti-semitism, its anti-intelligentsia attitude, and its internal bickering. These are all legitimate and worthwhile topics to include and should be discussed as he does very fully. It is just that ethnic studies frequently deal with the negative rather then the positive sides of immigrant life.

Like many fraternal histories I have read and scholarly papers which touch upon ethnic fraternals, Mr. Vassady dwells on the nationalistic, religious, and personal rivalries of the fraternal rather than such basic questions as the role of fraternals in unionizing workers and making them conscious of their condition; the effect they had in stabilizing ethnic communities by providing loans to individuals to buy homes or start businesses; the educational and socializing role of the fraternals; the development of ethnic consciousness among members; and many other issues dealing with the economic, social-political life of immigrants and other ethnics.

The paper only superficially deals with these topics. Mr. Vassady mentions that the Association touched nearly three hundred Hungarian-American settlements, but other than the national conventions, at which there seemed to be so much conflict, we don't really know much about the relationship of the Hazleton Verhovay Chapter with other chapters. What, if any, relationship did they have to the development of Hungarian nationalism prior to 1914 and after 1918?

Mr. Vassady's paper simply leaves too many questions unanswered, and we are left with a negative opinion of the Association and not an assessment of its accomplishments in so far as it relates to the Hungarian-American community. However, the study points up the importance of fraternals when studying ethnic groups. Too often fraternals, like the Verhovay, are ignored when studying the life of immigrants.

Chapter Four

CORPORATE ATTITUDES TOWARD LABOR ORGANIZATION
THE CONTROVERSY OVER THE PRICE OF POWDER IN THE LACKAWANNA VALLEY, 1888–1889

Perry K. Blatz

An appropriate introduction to corporate attitudes toward labor organization in the anthracite fields in the late nineteenth century can be found in the testimony of Ariovistus Pardee before a committee of the United States House of Representatives in February 1888. The committee had convened to investigate the conditions leading to the ongoing strike of mine workers in the southern anthracite regions of Lehigh and Schuylkill, and the concurrent sympathetic strike of some employees of the Reading Railroad. Pardee, one of the anthracite industry's pioneers, had helped open the Lehigh region to systematic exploitation in the 1830s by doing railroad surveys, and he began his own mining business in the region in 1840. By 1888, he had entered his seventy-eighth year,[1] and he and his sons operated some twelve collieries which produced more than seven hundred thousand tons of coal in the strike-shortened year of 1887.[2] Pardee not only opposed the mine workers' efforts to unionize but, more importantly, he viewed the very concept of a labor organization representing workers as illegitimate in and of itself. Since Pardee held that such organizations could in no way be genuinely representative, he saw them, at best, as irrelevant to his business, and he refused to have anything to do with them. Pardee made these points in characteristically blunt fashion in the following excerpt from his testimony, in which he was questioned by Representative William J. Stone (Democrat-Missouri).

> Q. Will you state, then, what objection you would have to conferring with your own men as Knights of Labor and as Amalgamated Association men . . . ?
> A. I claim they have no business between us and our men.
> Q. Do you claim that the men have no right to organize?

A. No; I do not care how many they organize, and they can belong to as many as they please.
Q. Then, if you have no objection to their organizing into labor organizations, what objection have you to conferring with them in that capacity?
A. Because I look upon the leaders of the organization as meddlers. It is not the miners themselves, but I look upon the leaders of these associations as mischievous rascals, stirring up mischief all through the country. Did you ever know a time when there was so many causeless strikes as since the Knights of Labor have obtained the power they have now?
Q. We do not want to argue the question, we want to hear your side.
A. This is part of my side.
Q. Now, I want to know if the men in fact did have cause to complain, would you object to confer with them?
A. No; not with our employees.
Q. Would you object to conferring with them, even if they came as representing some organized body?
A. I would not deal with them at all.
Q. Suppose the employees of these mines, in their various capacities, should organize; would you refuse to confer with them in that capacity?
A. I am not dealing with any organization or leaders of any organization.
. . .
Q. Well, suppose the association known as the Miners' Amalgamated Association, composed exclusively of your miners, should send representatives to you in behalf of the men, would you treat with them?
A. If they would come as our employees I would treat with them, but as representing an association I would not.
Q. Why not?
A. Simply because I would not. I have been on a chronic fight with these organizations. I commenced it in 1847.
Q. You are opposed to these organizations?
A. I am opposed to their interfering between me and my employees.
Q. But they are your employees although they may belong to an organization. They come to treat with you about matters between you and them.
A. If they come as a committee of employees I would treat with them; if they come as a committee of an organization I would not treat with them.
Q. Then before you will treat with them they must come in their individual capacity?
A. In their individual capacity or as a committee of our own men and not as a committee of any association.
Q. Now, really, do you not think this is rather arbitrary?
A. It may be; perhaps I am an arbitrary man.[3]

While Pardee was surely one of the anthracite region's flintiest individual operators his attitudes toward labor organization basically were shared by the managers hired in the service of corporate capital to oversee the anthracite interests of the several railroads which dominated the production and distribution of anthracite coal. Two of the most prominent of these managers were Samuel Sloan, President of the Delaware, Lackawanna and Western Railroad (DL&W), and William R. Storrs, the manager of the railroad's Coal Department, located in Scranton. When Sloan learned in February 1888 that mine workers in the Wyoming region, to the north of the Schuylkill and Lehigh regions, might make demands for wage increases similar to those which had precipitated the current strike by their southern brethren, he expressed his support for Pardee and other operators who had refused the strikers' demands. In a letter to Storrs, Sloan endorsed the actions taken by

> Mr. Pardee who with Mr. Coxe [Eckley B. Coxe, another prominent operator from the Lehigh region] first resisted the demands and then again Mr. Corbin of the Reading [Austin Corbin, president of the Reading Railroad] who in so manly a manner has resisted dictation from the order [Knights of Labor] as to their business and work.

Any hint of acquiescence to demands from the men of the Delaware, Lackawanna and Western would "throw away all *they* [Pardee, Coxe, and Corbin] have endeavored to do for *capital* (if I may use the term and I cannot help it.)"[4] Of course, Storrs found himself in complete agreement with his boss, and he stated that "it may be not only right but vital, for Capital to stand . . . in defense of its interests and to limit, if not defeat the power of an organization [Knights of Labor] whose avowed object is to control, if not crush capital." Storrs could not resist the temptation to indulge in some philosophizing, so he added that "next to life, and its defense, the right to possess capital and enjoy property is the dearest."[5]

It certainly should come as no surprise that men such as Storrs, Sloan, and Pardee opposed labor organization. However, in order to understand the depth and character of that opposition, which would in 1902 precipitate one of the nation's most momentous episodes of labor protest, it is necessary to examine in detail corporate attitudes and the way in which they were expressed. Such is the purpose of this paper. If we had the opportunity to ask the men who ran the anthracite corporations what principle inspired their struggle against unionization, they would most likely respond that they fought in defense of economic freedom or, as Storrs put it, "the right to possess capital and enjoy property." Yet, as Rowland Berthoff has shown so skillfully, the freedom they sought, when applied to labor relations, was a "freedom to control," a desire for unbridled power over their employees.[6] Businessmen sometimes justified their quest for this freedom on pragmatic grounds, maintaining that they needed such control over their workers if their businesses were to prosper. However, when they tried to justify their need for such power to their workers, or discussed it with their fellow businessmen, they often did so in terms of the oldest model for labor relations, paternalism. Paternalism is a term often used in a wide variety of historical works, so it is necessary to specify the sense in which it is used here. The emphasis in this paper will not be that of Eugene Genovese, who brought to light the complex network of mutual bonds that arose between slaves and slaveholders in the antebellum South.[7] Rather, paternalism in this paper is intended to carry the emphasis given it by Richard Sennett, who stressed the severe limitations inherent in the concern that a paternal authority figure professes to have for his metaphorical children. Sennett viewed paternalism as a "false love," in which an authority figure promises a paternal sort of care for his subjects, but insists on exercising that care entirely on his own terms, without consulting the recipients of his benevolence. The falseness of the authority figure's professions of love quickly comes to the surface whenever his authority is challenged. While a truly loving father figure would tolerate misbehavior from his children, encourage their desire for independence, and allow them the freedom to mature and leave childhood behind, the paternalistic authority figure will only care for his metaphorical children as long as they obey his rules and show no inclination to contest his control.[8]

Employers have justified paternalism by asserting that workers were childlike in many ways, and could prosper only under the boss's guiding hand.[9] They saw no logical inconsistency in denying the legitimacy of workers' organizations at the same time that they busily worked together to apportion the market among themselves or to decide how much they would pay their employees. They saw their cooperative efforts as genuine, but they believed that such efforts by their workers were utterly illegitimate, inasmuch as the workers supposedly lacked the maturity necessary to look after their own interests. Of course, the employers' thought process was by no means a rational one. Its primary purpose was self-justification and its pursuit of that goal was marked by recurring instances of wishful thinking. Contradictions abounded throughout the entire process, as the very employers who bemoaned their workers' supposed inability to know their own minds were quick to praise the workers' "manliness" whenever they sided with management. The intricacies of the thought process can be examined in detail in the correspondence of the managers of the DL&W during the closing months of 1888 and the opening months of 1889, when the company's miners organized to demand a cut in the price they paid the company for blasting powder, a major expense for each miner.

The price of powder fluctuated greatly during the post-Civil War years;[10] and this caused the companies in the Wyoming region, after consulting committees of their miners, to arrange for the price of powder to rise or fall along with wages, with an upper limit of three dollars for each twenty-five pound keg. An advance of wages in the Wyoming region in 1880 brought powder to this limit; and the price remained three dollars per keg when the Wyoming companies granted a further advance in 1882.[11] At the end of the unsuccessful strike of 1887–1888, powder still stood at or near three dollars throughout the anthracite fields.[12] Nevertheless, the fact that the price of powder had not changed for the miners gave them little cause for celebration because the price that the coal companies paid the powder companies had dropped substantially in the meantime. While the DL&W paid $2.20 per keg for powder in 1879, by 1890 it paid only $1.50 per keg.[13] The miners were further chagrined that, as the companies mined more intensively, miners often had to work more difficult veins, and to do this they had to use more powder. Of course, with powder becoming a more and more lucrative part of their business,[14] the coal companies had little incentive to heed calls for a decrease in the price they charged the miners for powder.

Soon after the strike of 1887–1888 in the southern regions ended in defeat for the mine workers, the Reading Railroad announced that it would reduce the price of powder to $1.50 per keg.[15] The Reading insisted that its miners had suffered no real injustice under the old price, since the company had taken the price of powder into account in setting rates for mining. Nonetheless, "in order that there should be no apparent cause for complaint," the Reading reduced powder and other mining supplies to their approximate cost to the company. Although the Reading claimed that its action increased miners' earnings,[16] others reported that the company decreased the rates paid to miners accordingly, so that any advance was minimal.[17] In any event, W. R. Storrs of the DL&W expressed his discontent over the Reading's action, stating that the Reading's reduction "would force on us sooner or later the question of revision—an advance [of wages] direct or indirect by less price for supplies." He even feared the possibility of a strike over the issue.[18]

The Knights of Labor had failed to attract substantial support in February 1888 for their effort to persuade the northern region's mine workers to join the ill-fated strikers to the south, so the union needed some time to generate interest in the powder question in the Wyoming region. In October, Storrs told President Sloan about a meeting of the Knights planned for the upcoming weekend to consider the price of powder. Storrs evidently feared that the Knights would attract quite a following, for he mentioned that a "voluntary advance" might stop the Knights. He added that he had told "our men last year when we could we would [advance wages], without their asking."[19] However, his concern soon abated when he heard that the meeting had been a failure.[20]

Nevertheless, the Knights persisted. On November 12, Assistant General Coal Agent William H. Storrs, minding the Coal Department while his father was out of town, reported to Sloan that each of the three leading companies in the Scranton area—the DL&W, the Delaware & Hudson (D&H), and the Pennsylvania Coal Company—had received notice that a committee of miners wanted to meet with the local mining manager for each company.[21] When the younger Storrs met the committee, he told them to talk with his father when he returned. Then the committee went to see A. H. Vandling, manager of the D&H's coal operations. The members of the committee requested a reduction in the price of powder to two dollars per keg. Vandling gave them no encouragement on that score, and counterattacked by questioning them sharply as to whether they were D&H employees or a committee representing the Knights of Labor. They replied that they were both.[22] This meeting accomplished nothing, inasmuch as Vandling and the junior Storrs believed that the committee in no way "represented either the majority of DL&W or D&H employees."[23]

The powder question grew much more complicated several days later when Robert M. Olyphant, president of the D&H, privately expressed himself in favor of a cut in the price of powder. President Sloan, of the DL&W, wrote W. F. Halstead, head of the company's railroad operations in Scranton, that he was "very much annoyed by Mr. Olyphant."[24] The younger Storrs reported that Vandling too seemed to disagree with Olyphant; and that Vandling feared that the Knights' Committee would receive credit for any reduction in powder at this time, thereby "opening a door that would be very difficult to close." Apparently to appease Sloan, Olyphant had promised not to act without consulting him and the senior Storrs.[25] Sloan then advised W. R. Storrs to talk with the managers of the other coal companies, but urged him to do so "with great secrecy and caution" so word of the proposed reduction did not reach the public.[26] To mobilize the companies in the Wilkes-Barre area against Olyphant's plan, the elder Storrs met with the managers of the Lehigh and Wilkes-Barre Coal Company, the Lehigh Valley Coal Company, and the Susquehanna Coal Company. They had refused to negotiate with the Knights, and agreed to oppose any reduction in the price of powder. In addition, all concurred that any reduction in powder would have to be accompanied by a corresponding cut in the rates paid miners per car or ton. W. R. Storrs felt confident that, faced with a united front, the D&H would change its position —"I think [they] will cooperate. They are not likely to pay six cents per ton more [for labor] than others."[27]

Regardless of what the D&H decided to do, W. R. Storrs could not ignore the Knights of Labor. He expressed his attitude toward that labor organization in rather stark terms in a memo he wrote in early December. Referring to the existing arrangement on powder as a "contract" between the miners and the company, he maintained that "only part of the men" desired a reduction and those had been "put up to it by Knights of Labor who ask the reduction to get credit by their organization."[28] Fear of the Knights and contempt for them mingled simultaneously in Storrs' thinking, and his fear can be seen in the following sentiments which he had expressed earlier, in February 1888.

> No one outside can tell what a secret organization will do—only the managers [of the organization] know. While they desire to be respectable and orderly the better to delude friends and foes, their purposes must be secured at any cost. Their followers do what they are told and rely on what is promised.[29]

In his discussion in December 1888 of the Knights' effort to reduce powder, he stated that success in this endeavor would only whet the Knights' appetite for further conquests, especially state legislation "imposing unreasonable conditions, liabilities and obligations upon the Company's operations."[30] In perfect harmony with these views was Storrs' belief that the mine workers, except for a few "agitators," were satisfied with their conditions, since the DL&W paid just as much as any other mining concern in the anthracite fields, and more than other local industries.[31] However, Storrs' sanguine assessment of the labor situation may have been based less on fact than on his desire to ignore, for the time being, the heavy toll that would be levied on the DL&W's profits if the company met the Knights' demands.

Meanwhile, Sloan stepped up his pressure on Olyphant and managed to persuade him to delay any action on powder until 1 March 1889. The DL&W's president also hoped that unity on the part of the managers in Scranton and Wilkes-Barre might lead Olyphant to order a further postponement.[32] However, as trouble from the D&H lessened, W. R. Storrs had to face increased pressure from the workers, as a committee of DL&W employees asked to meet with him on December 27.[33] The Knights then busied themselves in efforts to organize the miners and to select delegations from each of the companies to meet with each manager.[34] Clearly the Knights wanted to avoid giving W. R. Storrs and his counterparts a pretext for refusing to talk with the committees, on grounds that the committees weren't comprised solely of their own men.

The way in which W. R. Storrs conducted labor relations is revealed in his description of his meeting with the committee of DL&W workers. He first interrogated them concerning whether or not they worked for the company. Apparently satisfied on that score, he appealed to them to drop their demand for a reduction of powder. He emphasized the sanctity of the rather murky agreement that the company had made concerning powder with a committee of miners many years ago. He expressed "surprise" that anyone would want to alter that agreement; however, "if a good majority desired a new deal on powder which involved a general readjustment, we would at a favorable time consider the matter." Not surprisingly, he advised the workers that "the present was not . . . a favorable time and *we would not take up the matter.*" When the delegation told him that the other companies were willing to consider a reduction of powder if the DL&W would, Storrs responded innocently. "I said that is strange—others, act in other things independent and I do not believe they are willing to do what we do. It is possible, but not at all probable."[35] Of course, Storrs did not make any mention of his untiring efforts to unite his fellow mining executives against Olyphant's proposal to reduce the price of powder.

Storrs then took the offensive and chastised the men, telling them he knew "that you have been set up to this move by outside influences—and would not take this step except for this pressure." He informed them that they could "not afford to disturb" the present wage scale which had "worked so satisfactorily so long for . . . the men and for the company." Holding out the promise of paternal care, he told them of President Sloan's desire "to do all . . . [he] could fairly for the men," and he counseled them "to be content." He closed with an obvious parting shot at the Knights, telling the committeemen that he "thought them fully competent to look after their own affairs."[36] Not satisfied by the meeting, the committee then tried to go over Storrs's head by presenting their case in writing to Sloan in New York, but Sloan merely sent their letter to Storrs, asking Storrs to tell them that "the President would not and does not refuse to hear them," but wished that they deal with Storrs inasmuch as the matter was under his jurisdiction. Sloan added "we must be firm but be fair with the men *avoid any issue!*"[37]

In discussing the outlook for the immediate future, Storrs wrote Sloan that his general inside superintendent, Benjamin Hughes, saw no chance of a strike for several months. Hughes had also reported that those who had been joining the Knights and were responsible for pressing the issue were "younger men." Storrs doubted that a refusal to lower the price of powder would provoke a strike, "yet it may. If it does, the present is [a] favorable time for us, and not so for the men—better now than in the busy season."[38] Indeed, many large orders for the winter were customarily filled before the first of the year, and miners often experienced their greatest amount of idle time during the first quarter.[39] Nonetheless, the DL&W had to consider the effect of the D&H's proposed cut in the price of powder. If that occurred, not only would all other companies face almost unbearable pressure to follow suit, but mine workers other than miners would want a corresponding increase. Taking all this into account, Storrs wavered in his resolve and told Sloan that it might be advisable to post a general notice on January 1, "outside of the [Knights] Committee," announcing an increase of five per cent to non-miners and approximately the same amount to miners through a reduction of $1.00 or $1.25 in the price of powder. Storrs went on to detail for his boss the many factors militating against such a course of action, including the fact that the men "have no just or equitable claim for this now." While Storrs "would only think of it [an advance] as a compromise to avoid [a] strike," his counterparts at the other companies either preferred to risk a strike or doubted the possibility of a strike even if they continued to refuse the Knights' demands.[40] Sloan decided to hold firm against any advance and wrote Storrs, "I do not think you will have any trouble." However, he also warned that "the Knights will do their best to exert influence and must be watched."[41]

There was little change in the next few weeks; but on January 24, Olyphant met Sloan and told him he was now considering a reduction of only fifty cents per keg, and that he still considered the entire matter open for further discussion. Nonetheless, he was resolved to make some reduction of powder on March 1.[42] Storrs expressed regret that Olyphant "does not see his own interests" concerning powder. He went on to analyze powder's role in the wage question, and stated that, contrary to public perception, "there is no wrong done the miner in the matter now." To reduce powder was merely to advance the miner's wages, and it should be done only "when we get [a] corresponding reduction of his contract prices." In such a scheme, of course, the cost of mining to the company would change little. Harkening back to an arrangement perhaps more common in bygone days, Storrs expanded on the injustice he saw in the miners' demand.

> Men are sometimes employed at say twenty dollars a month and board or paid forty dollars a month and board themselves. The miners expenses . . . Powder and other supplies are the board which he pays for and we cover in the price paid him. He asks us to advance his wages by paying part of his board.[43]

His backbone now stiffened. Storrs believed that, even if the miners threatened to strike, the company should refuse their demands and "take the consequences."[44]

With the March 1 deadline fast approaching, Sloan did his best to dissuade Olyphant from taking action on powder.[45] He managed to get Olyphant to consider a reduction of contract rates to miners to correspond with a reduction of powder. Heartened by this development, Storrs told Sloan that he would confer with representatives from other companies to encourage joint action along that line. Once again turning to analyze the controversy over powder, Storrs labeled two culprits—the public and the Knights.[46] Editorials had pointed to the disparity between what the companies paid for powder and what they charged the miners, and they usually closed by calling for a cut in powder.[47] Of course, as far as Storrs was concerned, the public lacked "any correct knowledge of the facts or contracts." More important was the role played by the Knights of Labor, who were "agitating to make capital for their organization." In Storrs's view, the company could not afford to yield an inch; for if the Knights were

> recognized and granted this demand, it will be followed by others, in harmony with the legislation which they are trying to secure at Harrisburg, hostile to Companies, operators, and their agents by imposition of unreasonable liabilities and obligations.

Storrs concurred in Olyphant's view that it would be best to "remove the powder question" from the wage picture at a favorable time; but it must not be done by assuming "serious burdens—in cost of coal, and outside interference."[48] In his response, Sloan pointed to the major problem for the company—the stand taken by the D&H. "If D&H were with us I would feel more easy but they are *not*." He feared that the D&H aimed to isolate the DL&W so that the D&H would receive credit for any reduction, while the DL&W would be blamed if no reduction occurred.[49]

On January 31, Sloan seemed optimistic, and told Storrs that "the Knights are seeking position everywhere! here they will I think not succeed."[50] However, perhaps because he feared Olyphant would carry out the reduction, he wrote Storrs the next day and told him that Olyphant "has placed DL&W in the gap . . . it is a great pity we should do anything at all."[51] In his response, Storrs seemed almost resigned to the reduction; and he advised Sloan that, if the D&H insisted on reducing powder on March 1, the DL&W should reduce on February 15, "in advance of the time when the men will look for it." Still, if the company could blunt the Knights' thrust now, "and general dullness [in the coal trade] prevails, . . . [any reduction in powder] may be delayed for a time."[52] Obviously, Storrs realized that a miner who had little opportunity to work would be less likely to quibble over the price of powder than one who had been working full time.

Sloan, however, had not yet given up hope of re-establishing corporate solidarity. He sent his first vice-president, E. R. Holden, to meet with Edwin H. Mead, president of the Pennsylvania Coal Company, the third-largest anthracite producer in the Lackawanna Valley. Mead believed that Olyphant was having second thoughts about "agitating the matter (this is private) and would be glad to see his way out of it."[53] On another front, Storrs reported to Sloan on February 8 that he had discussed the powder question with Thomas H. Phillips, general manager of the Lehigh and Wilkes-Barre Coal Company's collieries in the Wilkes-Barre area. Phillips favored a reduction of powder to $1.75 per keg along with a "corresponding reduction in the price per car paid the miners." Of course, "concert of action by all the Companies" was the key to such a plan; and the managers of the Lehigh Valley Coal Company and the

Susquehanna Coal Company had also assented to it. The D&H remained the major obstacle.[54]

Sloan could not get Olyphant to agree to leave powder alone, but Olyphant did express a willingness to reduce both the price of powder and the miners' price per car. Olyphant suggested that the local managers confer concerning such a step. Sloan thought that a conference would be helpful, but at the same time he saw pitfalls in reducing both powder and miners' rates. He wrote Storrs, "I can see only one way and that is follow D&H" in Olyphant's original proposal to cut only the price of powder. He added, "I do not think you can or ought to reduce wages."[55] In any event, Storrs met with the other local managers and reported to Sloan on what had transpired. All had agreed that they preferred not to change the price of powder. If this proved impossible, they hoped to reduce the miners' rates per car as well as the price of powder, "so he [miner] gets the same as now." Nevertheless, if powder was reduced and miners' rates were not adjusted, those workers receiving a daily wage must be advanced an approximately equal amount—"in other words treat all alike."[56]

A meeting of company presidents and their mining managers held in New York City on February 18 succeeded in cementing the united corporate front. At that meeting, the presidents and managers agreed on a plan which would allow the miners to vote as to whether they preferred no change at all in mining rates or powder, or a reduction in mining rates coupled with a reduction in powder. Miners would not be permitted the option of voting for the Knights' demand, a lower rate for powder with mining rates held steady.[57] Yet, even with the alternative so carefully rigged, the bosses were still fearful. Sloan wrote Storrs that "it is a very great pity you have to bring the matter we talked of yesterday before the men at all."[58] Storrs agreed that "it is a difficult matter." The vote had to be planned very carefully. Although the miners' choices were limited to the options the companies were prepared to accept, Storrs warned nevertheless "we must act simultaneously through all the mines. Some men may leak, which would defeat the plan." Since only a few of the DL&W mines were operating currently, "we must arrange for all to work the day of the canvas." The foremen would conduct the vote; and they "must be instructed and some must have assistants," because it would be impossible in some of the larger mines for one man to ask each and every miner for his "vote" during one working day. General Inside Superintendent Hughes had assured Storrs that "a large majority will prefer no change," but due to the Knights' supposedly tyrannical hold the miners "may not be willing to express themselves." Storrs feared that if the miners voted for change in their contract rates and the price of powder, they could "construe our movement [the vote] as weakness, and may demand reduction of Powder without corresponding reduction of mining [rates]." After the companies' crafty gesture toward industrial de-

mocracy, such a reduction in mining rates would "be more difficult to arrange
. . . then it would be to carry it on notice and without canvas."[59]

The paternalism which dominated Storrs' view of labor relations came through clearly when he explained why he agreed with Sloan that it was so unfortunate to hold the vote. "Absolute silence, which they [workers] construe as strength, or bold decisive action should govern us—not discussion."[60] In other words, companies should confront workers with straightforward edicts, and not allow workers any role, no matter how truncated, in the decision-making process. For Storrs and Sloan, the best policy in labor relations was to treat workers like children—unruly children who could not be trusted. From time to time the company acted in a somewhat fatherly manner toward them, but certainly not with the fostering care to which Storrs had alluded when he had told the miners' committee of President Sloan's constant concern for the men. The company wanted to decide all questions related to labor on its own, without allowing the mine workers any role other than to choose individually whether or not to work for whatever wages the company offered. Even the vote, so carefully set up by the companies to obtain the desired result, was viewed by Storrs and Sloan as a potentially fatal concession, leading to the loss of discipline and control. Time and again, in justifying their actions, companies like the DL&W insisted that they needed the freedom to control their workers, not only for the sake of their businesses, but for the sake of the workers as well. The workers could never be allowed a larger, more mature role in decision making, for in the view of the managers, they simply lacked the capacity for it.[61] Most importantly, they could never know their own minds. They were notoriously susceptible to the blandishments of agitators like the Knights, who would say what men like Storrs and Sloan could not bear to hear, that the managers did not have the workers' best interests at heart.

The managers scheduled the vote for Tuesday, February 26;[62] the workers were informed on Monday that all collieries would operate the next day, and that all workers were expected to report for work. On Tuesday morning, soon after the miners entered their chambers, each one was confronted by a foreman, fire boss, or member of the mine's surveying corps, who asked the following question: "Do you desire a reduction in the price of powder, with a corresponding reduction on the price paid on a car, or are you satisfied to have prices remain as they are?" As George S. Boyle, an organizer for the Knights of Labor, commented, "it will be observed that in the adroit manner of asking these questions any other than a negative answer could scarcely be expected."[63] This aptly described the result. The miners of the DL&W voted 1,224 to 128 against any change, and those of the D&H and Pennsylvania Coal Company voted similarly.[64]

The president of the DL&W congratulated Storrs on his success, writing that "your action is certainly most satisfactory!"[65] Storrs took the opportunity to make a detailed, self-serving interpretation of the vote. First, the company's victory demonstrated "the false position of the Knights." Second, since newspapers and much of the public had favored a reduction, the vote was "a proper rebuke" to the public "for meddling with a matter which did not concern them." Finally, in apparent reference to the old agreement to keep powder at three dollars per keg, Storrs stated "it would have been a mistake on our part to have broken faith with the men." He closed by telling Sloan that "the noes were largely Knights" and that "I feel we can stand more criticism with the men on our side and $200,000 a year saved."[66] Storrs was further heartened by reports the following day that several D&H miners who had originally favored a change had since asked to have their votes switched. He was also pleased to report that the men of the DL&W's Pyne Colliery, where the vote took place one day later, voted eighty-four against change and ten for change, with twelve abstaining—"and they were not taken by surprise." Storrs seemed serene, stating "we have the advice of the men and should be content."[67]

Storrs' analysis illustrates the extent to which wishful thinking dominated his thought process. In concocting the plan, he realized that coordination and surprise were essential—that this was no ordinary vote. Yet, once the results came in, he apparently managed to persuade himself that the vote truly represented the miners' views, and that to have reduced powder would have constituted a breaking of "faith" on the part of the company. Just a week before, Storrs had expressed great apprehension about allowing the workers to exercise even the very limited decision-making power permitted them in the vote. After he had obtained the desired result, however, he sounded as if he had been confident of the outcome; and now he was eager to treat the workers as trusted members of the DL&W family. Storrs wrote his boss that "this action of the men with their good behavior last year entitles them to favorable consideration when circumstances will warrant it."[68] Sloan agreed: "the men deserve and will have from us favorable consideration when [the] time comes!"[69] Yet, as it turned out, the DL&W's mine workers had to endure a very long wait before collecting their reward.

Faced with these shrewd corporate maneuvers, some workers tried to resist. Of the approximately 130 miners at Hallstead Colliery, 108 of them signed the following statement which Storrs received on March 5.

> We . . . do believe our Employers has used us unjust in the manner they have treated us regarding the Powder Question, therefore after due and carefull consideration we condemn the action taken by them. We did not understand the Bosses when they came around the mines to us, therefore we revoke our vote of the 26th . . . and vote as follows that we have the powder reduced from $3 to $2 per keg and all the wages paid the same as they stand at present no alterations in mining whatever.

Not surprisingly, Storrs attributed this action to the influence of a local leader of Knights who at a recent meeting had spoken a "tirade against the Company." To assure himself of the vote's legitimacy, or perhaps more importantly to assure President Sloan, Storrs informed Sloan that the inside foreman at Hallstead had reported to him that "no compulsion was used on the canvas." Whenever any miner asked for an explanation of the limited choice available, he was told that the company simply could not reduce powder without reducing the miners' rates as well.[70]

Early in 1893, the miners of the Lackawanna Valley received their only advance of any kind in the decade, as the Pennsylvania Coal Company, the D&H, and the DL&W cut twenty-five cents from the price of a keg to match the price paid for powder by miners in the Wilkes-Barre area, $2.75.[71] Miners obtained a more substantial reduction of powder only as part of the settlement of the United Mine Workers' six-week anthracite strike of 1900, after which the miners had to pay only $1.50 per keg.[72] Certainly, with regard to the price of powder, anthracite miners realized that they had gained little by waiting patiently for the companies to grant them relief. As more and more workers recognized the one-sided nature of paternalism and moved to wring concessions from the operators by collective action, most operators became only more convinced of the evil embodied in labor organization. They continued to oppose the union intensely, long after the United Mine Workers (UMW) succeeded in organizing the anthracite industry at the turn of the century. Their opposition resulted in the charade of operators refusing to allow any mention of the UMW in the agreements they signed with the union's officials in the anthracite region, so that the operators could continue to maintain that they had not recognized the union. By the time that they granted formal recognition to the UMW under federal pressure in 1920, so much bad blood had been generated that the recurring deadlocks of the 1920s between labor and management should come as no surprise.[73]

The controversy over the price of powder demonstrates the tenacity and shrewdness with which American businessmen struggled to preserve what they perceived as their inalienable right to run their businesses as they pleased and manage their workers as they saw fit. Even in the years after 1902, with the UMW well-established in anthracite, operators sought to maintain their control by showing a greater willingness to grant increases in wages and benefits than to agree to the union's demands for changes in work rules.[74] A similar tendency has been discerned by David Brody, who has detailed the doggedness with which American management has held onto and even expanded its sphere of control in the years after World War II, while simultaneously consenting to improve packages of wages and benefits for its employees.[75] Perhaps the human element in the industrial landscape has changed less than we might expect since the turn of the century. Strong echoes of paternalism persist today, with corporations seeking to comfort their employees by assuring them that they are not merely workers, but members of the corporate family.[76] Such expressions may merit little more than a cynical epithet from us, but their prevalence should make us reconsider their significance. As long as having a job is perceived as a privilege and not a right, employers will hope to substitute the false love of paternalism for the somewhat more secure rights embodied in a contract.

NOTES

[1]C. Pardee Foulke and William G. Foulke, *Calvin Pardee, 1841–1923: His Family and His Enterprises* (Philadelphia, 1979), 27–30.

[2]This is calculated from data in Pennsylvania, *Report of the Inspectors of Mines for the Anthracite Coal Regions of Pennsylvania for the Year 1887* (Harrisburg, 1888). In addition, the coal company begun by Pardee's brother-in-law, George B. Markle, whom Pardee had helped to enter the business, mined over 300,000 tons of anthracite in that year. For the family story see Foulke and Foulke, 38–39, 92–93.

[3]U. S. Congress, House Select Committee on Existing Labor Troubles in Pennsylvania, *Labor Troubles in the Anthracite Regions of Pennsylvania,* 1887–88, H. Report #4147, 50th Cong. 2nd sess., 1889, 554–55.

[4]Samuel Sloan to W. R. Storrs, 6 February 1888, Corporate Records of the Delaware, Lackawanna and Western Railroad, George Arents Research Library, Syracuse University, Syracuse, N.Y., hereafter cited as DL&W Records—Syracuse.

[5]W. R. Storrs to Sloan, 6 February 1888. Corporate Records of the Coal Department of the Delaware, Lackawanna and Western Railroad, Lackawanna Historical Society, Scranton, Pa., hereafter cited as DL&W Records—Scranton.

[6]Rowland Berthoff, "The 'Freedom to Control' in American Business History," in *A Festschrift for Frederick B. Artz,* eds. David H. Pinkney and Theodore Ropp (Durham, N.C., 1964), 158–80.

[7]Eugene Genovese, *Roll, Jordan, Roll: The World the Slaves Made* (New York, 1974).

[8]Richard Sennett, *Authority* (New York, 1980), 50–83.

[9]It is not possible here to make an exhaustive listing of historical works examining paternalism in relations between employers and their employees. However, a few of the most significant studies in this area are Sidney Pollard, "The Factory Village in the Industrial Revolution," *English Historical Review,* LXXIX (July 1964), 513–31; Anthony F. C. Wallace, *Rockdale: The Growth of an American Village in the early Industrial Revolution* (New York, 1978); Stanley Buder, *Pullman: An Experiment in Industrial Order and Community Planning, 1880–1930* (New York, 1967); Melton Alonza McLaurin, *Paternalism and Protest: Southern Cotton Mill Workers and Organized Labor, 1875–1905* (Westport, Conn., 1971); and Tamara K. Hareven and Randolph Langenbach, *Amoskeag: Life and Work in an American Factory City* (New York, 1978).

[10]Frank Julian Warne, "The Anthracite Coal Strike," *Annals of the American Academy of Political and Social Science,* XVII. (January 1901), 44.

[11]W. H. Storrs to W. H. Truesdale, 30 March 1899, DL&W Records—Scranton.

[12]Peter Roberts, *The Anthracite Coal Industry* (New York, 1901), 135. Some

companies in the Wilkes-Barre area charged only $2.75 per keg, at least as early as February 1888. See Parrish Coal Company Payroll, February 1888, Papers of the Parrish Coal Company, Wyoming Historical and Geological Society, Wilkes-Barre, Pa. As of April 1889, the Lehigh and Wilkes-Barre Coal Company also charged $2.75 per keg. See *Wilkes-Barre Record,* 4 April 1889.

[13]W. H. Storrs to W. H. Truesdale, 30 March 1899, DL&W Records—Scranton.

[14]Roberts, *Industry,* 134.

[15]*Ibid.,* 135.

[16]Philadelphia and Reading Railroad Company, *Report of the President and Managers of the Philadelphia and Reading Railroad Company for the Year Ending November 30, 1888, together with the Report of the Philadelphia and Reading Coal and Iron Company* (Philadelphia, 1889), 17, 61.

[17]Warne, "Anthracite Coal Strike," 45; Carroll D. Wright, "Report to the President on the Anthracite Coal Strike," *Bulletin of the Department of Labor,* 7 (November 1902), 1169. Statistics compiled from *Mine Inspectors' Reports—1888* support the likelihood of some readjustment in mining rates. In that year, 153,496 kegs of powder were used at the Reading mines; and the company claimed that its reduction of mining supply costs "has resulted in practically increasing the wages of the miners . . . over $110,000." See *Report of the Philadelphia and Reading Railroad—1888,* 17. Obviously, miners did not receive the full benefit of the decrease in powder of $1.50, for multiplying 153,496 by $1.50 yields $230,244. Of course, the $110,000 mentioned by the Reading must have included the decreased cost of other mining supplies, and might even have reflected merely an overall rise in mining costs, caused by factors far afield from the decrease in the company's income from selling miners their supplies.

[18]W. R. Storrs to Sloan, 27 March 1888, DL&W Records—Scranton.

[19]W. R. Storrs to Sloan, 4 October 1888, DL&W Records—Scranton.

[20]W. R. Storrs to Sloan, 9 October 1888, DL&W Records—Scranton.

[21]W. H. Storrs to Sloan, 12 November 1888, Nov. 13, 1888, DL&W Records—Scranton.

[22]W. H. Storrs to Sloan, 15 November 1888 (first), DL&W Records—Scranton.

[23]W. H. Storrs to Sloan, 15 November 1888 (second), DL&W Records—Scranton. W. H. Storrs implied that the same committee of Knights of Labor met with both him and Vandling, although he recognized that it did include some employees of the DL&W and the D&H. However, the *Wilkes-Barre Record,* 18 November 1888, stated that a committee of DL&W employees went to see Storrs and a committee of D&H employees went to see Vandling.

[24]Sloan to W. F. Hallstead (sic), 16 November 1888, DL&W Records—Syracuse.

[25]W. H. Storrs to Sloan, 17 November 1888, DL&W Records—Scranton.

[26]Sloan to W. R. Storrs, 1 December 1888, DL&W Records—Syracuse.

[27]W. R. Storrs to Sloan, 1 December 1888, DL&W Records—Scranton.

[28]W. R. Storrs memo, n.d.—but placed among letters of 7 December 1888, DL&W Records—Scranton.

[29]W. R. Storrs to Sloan, 6 February 1888, DL&W Records—Scranton.

[30]W. R. Storrs memo, n.d.—but placed among letters of 7 December 1888, DL&W Records—Scranton.

[31]W. R. Storrs to Sloan, 7 December 1888, DL&W Records—Scranton.

[32]Sloan to J. R. Maxwell, 10 December 1888; Sloan to W. R. Storrs, 10 December 1888, 11 December 1888, DL&W Records—Syracuse.

[33]*Wilkes-Barre Record,* 18 December 1888.

[34]W. R. Storrs to Sloan, 26 December 1888, DL&W Records—Scranton.

[35] W. R. Storrs to Sloan, 27 December 1888, DL&W Records—Scranton.
[36] *Ibid.* Cf. *The World,* New York, 28 December 1888, for an account of the meeting about which W. R. Storrs said contained "some truth and some falsehood" in W. R. Storrs to Sloan, 28 December 1888, DL&W Records—Scranton.
[37] Sloan to W. R. Storrs, 29 December 1888, 31 December 1888 (quotations are from the latter), DL&W Records—Scranton.
[38] W. R. Storrs to Sloan, 30 December 1888, DL&W Records—Scranton.
[39] Roberts, *Industry,* 125–26; H. M. Chance, *Report on the Mining Methods and Appliances Used in the Anthracite Coal Fields,* Vol. AC, *Second Geological Survey of Pennsylvania* (Harrisburg, 1883), 385.
[40] W. R. Storrs to Sloan, 30 December 1888, DL&W Records—Scranton.
[41] Sloan to W. R. Storrs, 3 January 1889, DL&W Records—Syracuse.
[42] Sloan to W. R. Storrs, 24 January 1889, DL&W Records—Syracuse.
[43] W. R. Storrs to Sloan, 25 January 1889 (first), DL&W Records—Scranton.
[44] W. R. Storrs to Sloan, 25 January 1889 (second), DL&W Records—Scranton.
[45] Sloan to W. R. Storrs, 26 January 1889; Sloan to Olyphant, 26 January 1889, 28 January 1889, DL&W Records—Syracuse.
[46] W. R. Storrs to Sloan, 29 January 1889, DL&W Records—Scranton.
[47] *Wilkes-Barre Record,* 20 November 1888, 11 December 1888.
[48] W. R. Storrs to Sloan, 29 January 1889, DL&W Records—Scranton.
[49] Sloan to W. R. Storrs, 29 January 1889, DL&W Records—Syracuse.
[50] Sloan to W. R. Storrs, 31 January 1889, DL&W Records—Syracuse.
[51] Sloan to W. R. Storrs, 1 February 1889, DL&W Records—Syracuse.
[52] W. R. Storrs to Sloan, 1 February 1889, DL&W Records—Scranton.
[53] Sloan to W. R. Storrs, 1 February 1889, DL&W Records—Syracuse.
[54] W. R. Storrs to Sloan, 8 February 1889, DL&W Records—Scranton.
[55] Sloan to W. R. Storrs, 12 February 1889, DL&W Records—Syracuse.
[56] W. R. Storrs to Sloan, 13 February 1889, DL&W Records—Scranton.
[57] W. R. Storrs to Sloan, 19 February 1889, DL&W Records—Scranton.
[58] Sloan to W. R. Storrs, 19 February 1889, DL&W Records—Syracuse.
[59] W. R. Storrs to Sloan, 19 February 1889, DL&W Records—Scranton.
[60] *Ibid.*
[61] Examples of such corporate thinking can be seen in Buder, 147–201; Sennett, 62–72; and David Brody, *Steelworkers in America: The Nonunion Era* (Cambridge, 1960; reprint ed., 1969), 240–41. Of course, one of the bestknown and most blatant paternalistic utterances was made during the anthracite strike of 1902, when George F. Baer, president of the Reading Railroad, stated that workers were best advised to leave decisions concerning their future in the hands of "the Christian men to whom God in his infinite wisdom has given the control of the property interests of the country." Quoted in Ray Ginger, *Age of Excess: The United States from 1877 to 1914* (New York, 1965), 244.
[62] W. R. Storrs to Sloan, 22 February 1889, DL&W Records—Scranton.
[63] *Wilkes-Barre Record,* 4 April 1889, George S. Boyle is referred to as an organizer in *Wilkes-Barre Record,* 10 December 1888.
[64] Sloan to W. R. Storrs, 28 February 1889, DL&W Records—Syracuse; W. R. Storrs to Sloan, 26 February 1889, 28 February 1889, DL&W Records—Scranton.
[65] Sloan to W. R. Storrs, 27 February 1889, DL&W Records—Syracuse.
[66] W. R. Storrs to Sloan, 26 February 1889, DL&W Records—Scranton.
[67] W. R. Storrs to Sloan, 28 February 1889, DL&W Records—Scranton.
[68] *Ibid.*

⁶⁹Sloan to W. R. Storrs, 1 March 1889, DL&W Records—Syracuse.
⁷⁰W. R. Storrs to Sloan, 5 March 1889, DL&W Records—Scranton. For the number of miners at Hallstead see *Mine Inspectors' Reports—1888,* 79.
⁷¹W. R. Storrs to Sloan, 13 January 1893, DL&W Records—Scranton; Sloan to W. R. Storrs, 14 January 1893, 17 January 1893, DL&W Records—Syracuse.
⁷²Warne, "Anthracite Coal Strike," 49–50.
⁷³Rev. William J. Walsh, *The United Mine Workers of America as an Economic and Social Force in the Anthracite Territory* (Washington, D.C., 1931), 127–33.
⁷⁴For example, from 1902 into the 1920s, the operators repeatedly refused the UMW's demand that miners be paid for each ton of coal they mined, rather than being paid a certain sum for each car of coal. However, the operators agreed to increase wages in 1909, 1912, 1916, and 1920. See U.S., Department of Labor, Bureau of Labor Statistics, *Bulletin of the United States Bureau of Labor Statistics,* Whole No. 191, "Collective Bargaining in the Anthracite Coal Industry," by Edgar Sydenstricker, March 1916, 30; U. S., Department of Labor, Bureau of Labor Statistics, *Monthly Labor Review* XI (October 1920), 98; Walsh, 138–42.
⁷⁵David Brody, "The Uses of Power I: Industrial Battleground," in *Workers in Industrial America: Essays on the Twentieth-Century Struggle* (New York, 1980), 173–214.
⁷⁶In corporate publications directed toward former and current employees, there are frequent references to the corporation's resemblance to the family. The Goodyear Tire and Rubber Company's newsletter for its Akron employees is entitled *The Wing Foot Clan. News about Thee and Me,* 7 (June/July 1982), published by the Quaker Oats Company, contains an interview with Chairman of the Board Robert D. Stuart, Jr. When asked how the company has changed since the 1940s, he remarks that even though the scope of the business has expanded considerably, the company still has manageed to keep "the feeling of belonging to the Quaker family." In his opinion, "the sense of family and common purpose" at the company is remarkable. Corporate managers use similar appeals to discourage unionization. The southern regional director for the Textile Workers Union of America, Scott Hyman, had the following comment on one of plant manager Frank Urtz's ploys to combat efforts to unionize the Oneita Knitting Mills in South Carolina in the early 1970s. "Urtz was a smart man and some of his tactics were not stupid. He used these things that we may think are silly, like the analogy of the family, 'this was all family.' Well that happens to be a pretty doggone effective tactic." Quoted in Carolyn Ashbaugh and Dan McCurry, "On the Line at Oneita," *Southern Exposure* IV (Nos. 1 and 2), 33.

Chapter Five

HENDRICK B. WRIGHT:
"A PRACTICAL TREATISE ON LABOR"

James P. Rodechko

In the years after the Civil War, Pennsylvania politics, like politics on the national level, emphasized pursuit of office rather than public service. Politicians were concerned with personal success and looked upon political parties as instruments that allowed them to pursue that success.[1] Even as industrialism and its attendant problems, depression and labor unrest, emerged full-blown by the 1870s, politicians and the parties that they represented were more concerned to use the new issues to their own advantage than to offer workable solutions. Name calling, exaggerated accusations, and outright lies characterized the political process. Republicans contented themselves with waving the bloody shirt while Pennsylvania Democrats generally focused their attack upon Republican corruption. Individual politicians adjusted their views to accommodate any particular local issue that might gain popular support.

Hendrick B. Wright, a Wilkes-Barre politician who attained a significant reputation in Pennsylvania, appears representative of this political trend. For roughly fifty years Wright was a constant seeker after political office. As a Democrat whose political base lay within Luzerne County, he advocated labor interests and carefully cultivated poor miners who composed a large part of his constituency. Over the same period, however, Wright was a successful lawyer who served well-to-do clients, notably coal mining interests and railroad corporations. In addition, Wright was involved in a number of business ventures, most notably as organizer and president of the Wilkes-Barre Water Company and as a director of the Second National Bank of Wilkes-Barre. He speculated heavily in coal lands and eventually achieved financial independence through the sale of these lands.[2] The dichotomy between Wright's political stance and his business and financial background suggests that he was little more than an opportunist. In fact, the *Dictionary of American Biography* notes that his "oratory and facile pen" gave "him a deserved but unenviable title: as the " 'Old-Man-Not-Afraid-to-be-Called-Demagogue.' "[3]

Wright was born on 24 April 1808 in Plymouth, Pennsylvania. His father, Joseph Wright, was a farmer and merchant who traced his English ancestry in America back to 1681. Hendrick Wright began life in a primarily small town environment in which his family played a prominent role. As a young boy, he worked on his father's farm and learned to value education. He originally attended local public schools and in 1824 entered the Wilkes-Barre Academy. In 1829, he attended Dickinson College but withdrew after two years. By general agreement, he did not distinguish himself as a student. Nevertheless, in 1831 he returned to Wilkes-Barre to study law in the offices of John N. Conyngham. He was subsequently admitted to the bar on November 8 of that year. In later years he established a reputation as a capable lawyer, but one known for his ability to win cases through ridicule rather than hard evidence. One of his colleagues, Judge Stanley Woodward, remarked that Wright "laughed more cases out of court than the average lawyer won after most careful preparation."[4]

As a young lawyer, Wright turned his attention to politics. By the early 1830s he not only identified himself as a Jacksonian Democrat, but began to derive certain benefits from his party affiliation. In 1834, he was appointed district attorney for Luzerne County by George M. Dallas, a prominent Pennsylvania Democrat who later became vice-president of the United States. In 1835, Wright was elected and commissioned colonel in the Wyoming Volunteer Regiment of the Pennsylvania militia. Beginning in 1841, he won election to three successive terms in the lower house of the State legislature, gaining the post as speaker in 1843. In 1844, he was prominent enough to secure the chairmanship of the national Democratic convention, which met in Baltimore. There he worked successfully for the nomination of James K. Polk for president. In 1850, 1852, and 1854, Wright was a candidate for Congress, but won only once, in 1852. He was elected to Congress again in 1860 and actively opposed dissolution of the Union.[5]

The Civil War brought changes in Pennsylvania's political climate. The Democratic party, which had dominated the state for two decades, was now commonly identified as the party for treason and rebellion. In the decades that followed, Civil War issues often secured Republican victories while Democratic triumphs were occasional at best. When Democrats did seem strong enough to offset the "bloody shirt," they were usually so badly divided among various State leaders that they achieved little more than temporary success. By the mid 1870s, the State Democratic party was especially plagued by the in-fighting between William A. Wallace and Samuel Randall, two rising leaders who commanded substantial factions. Wright, like other State Democrats, had to evaluate his own chances according to the factional strife, and the new role his party confronted.

Changing political conditions were accompanied by new economic conditions. Beginning in the 1830s, anthracite coal mining became increasingly important in eastern Pennsylvania, especially in Wright's home-base in Luzerne County. There was particularly notable growth during the 1860s when anthracite production increased by 100 per cent. And as the tonnage increased, an ever larger number of workers entered the region to mine the coal. Many of the newcomers were foreign-born and principally of Irish birth. These newcomers constituted a powerful political force, and one largely left untended by existing political parties and the platforms they presented. Wright's own Democratic party preferred to hold to the theories of Jefferson and Jackson and accentuate its conservatism. It made little effort to deal with rapidly changing conditions or to appeal to the new groups of people who worked in the region.[6]

The postwar situation that the Democratic party confronted, along with emerging urban-industrial conditions in Wilkes-Barre and the Wyoming Valley, prompted Wright to direct increasing attention toward organized labor. He had been identified with labor interests before the Civil War but became more vigorous in his support as industrialism expanded.[7] In 1869, he published a series of articles in the *Anthracite Monitor* under the name "Vindicator." By 1871, Wright updated the articles to account for more recent trends and published them in book form. He now publicly identified himself as the author and titled the collection *A Practical Treatise on Labor.* The book offered comments on specific conditions in the coal mines, pointed to outstanding problems labor would have to overcome, and proposed a variety of solutions, many of which were vaguely stated and unlikely to achieve positive results.

A brief perusal of the *Practical Treatise,* as well as a realization of the author's political and financial background, suggests that Wright's sympathy for labor was nothing more than simple opportunism. In fact, it is rather commonplace to conclude that the book "was an obvious bid for labor support" and was published for the sole purpose of enhancing the author's political career.[8] The language Wright used and the glowing tribute that he heaped upon laborers indicated an unmistakable effort to ingratiate himself with the working classes. Laborers were informed again and again that the greatness of America resulted from their effort and diligence. They learned that they should not doubt themselves because neither books nor the possession of wealth conferred brains. Wright glorified the laboring man's shrewd common sense and claimed that "theoretical, compared with practical knowledge amounts to but little."[9] He claimed that laborers in America held an "exalted status" and constituted collectively the "king, lords, and commons." Whereas the English nobility numbered only a few thousand, Wright argued that American laborers composed a nobility of forty million.[10] Wright even tried to identify his own career with labor. He cited the hard and long hours he worked on his father's farm and claimed that he drew life-long benefits from this manual labor.[11]

The *Practical Treatise* played upon popular prejudices that Wright identified with the laboring classes. The author lampooned the manners and cultural values of the upper classes and praised the virtues of a simple life. He offered openly racist views that were compatible with emerging labor attitudes toward poor immigrants. Since Chinese laborers were immigrating to California in large number and were supposedly driving down wage levels because of a willingness to work cheaply, Wright called upon the government to halt Chinese immigration. He held that the Chinese were an inferior and "pagan" people who would drag down the high quality of American life and labor. At the same time, Wright carefully avoided negative comments about European immigrants who were becoming a prominent part of the coal-mining industry. Although many native Americans condemned the Irish as an inferior racial group, Wright cultivated them as potential political constituents in Luzerne County. He likened the Irish to Anglo-Saxon Americans as hard-working contributors to American life.[12]

The book helped to provide Wright with a well-defined reputation as a labor advocate in Luzerne County and throughout Pennsylvania. Irish-Americans in the coal region saw him as their defender and sought him out to speak at social gatherings and public meetings. In fact, even Irish-American leaders in New York City and other urban areas considered him a principal spokesman for Irish-Americans in the coal region. In the early 1870s, Democratic politicians across Pennsylvania looked to Wright as the party's labor spokesman and requested his advice about how best to deal with the labor vote. When he died on September 2, 1881, New York's *Irish World and American Industrial Liberator,* a newspaper that was generally identified with radical labor agitation, eulogized him as a long-time friend of Irishmen and labor's cause.[13]

Because of this reputation, Wright attained a fairly significant political role in the 1870s. He was important in the intra-party warfare in Pennsylvania and supported the Democratic faction led by Samuel Randall. In 1875 and 1876, he chaired Democratic State conventions and generally worked for Randall's interests. He was mentioned as a possible Democratic candidate for governor and for the United States Senate on several occasions during the early 1870s, but was successful in gaining the party's nomination only in congressional contests. He lost congressional elections in 1872 and 1874 but won in 1876 and 1878. In the latter two contests he campaigned as a Democrat but was elected mainly because he gained support from the Greenback Labor party. In fact, many State Democrats complained that he was more clearly identified with the Greenback Labor party than with the Democratic party. In 1880, perhaps because he sensed that political opportunities on the national level were at hand, Wright severed his ties with the Democratic party and identified fully with the Greenback Labor party. For a time, he was considered a possible presidential candidate on that party's ticket and actually received 214 votes on the first ballot at the national convention in Chicago. When James B. Weaver was nominated instead on the second ballot, Wright made one last unsuccessful bid for Congress in the fall election.[14]

Wright's political interests were paramount in his thinking, and the *Practical Treatise on Labor* served those interests well. But it is too easy to consider Wright from one perspective alone. Although it is difficult to distinguish political hypocrisy from genuine concern, the *Practical Treatise* indicates a complex human being who reacted to a variety of social and economic conditions that bothered him. Even if the book failed to offer very workable solutions to labor problems, it nevertheless showed Wright's hopes and aspirations, not simply for himself, but for American society and the socio-economic class he represented. Beneath the flowery words and the exaggerated identification with labor, Wright revealed both the values that he cherished and the fears that disturbed him. He was a man afraid that his world was rapidly disintegrating due to the weight of its enormous problems. In fact, the *Practical Treatise,* as much as it was an appeal for labor support, was a warning to men of his own kind that something had to be done to save the system which had offered them opportunities to attain power and status.

Wright was very clearly a man who placed heavy emphasis on hard work and individual effort. In fact, his words suggest that hard work itself, rather than the type of work performed, created a broad unity among human beings. All hard-working people, therefore, were laborers of one sort or another. Looking back upon his own early experience, and discounting his family's prominence and influence, he believed that constant effort and personal ability had enabled him to work his way upward. Moreover, just as he construed his later efforts as a lawyer and a businessman as a form of labor which flowed logically from his work on his father's farm, he assumed that the same process was at work for others. He argued that ninety per cent of all successful merchants rose from the laboring ranks.[15] The merchant and lawyer held higher status, but hard work still bound them to those lower on the socioeconomic ladder. In a similar way, the skilled mechanic, the farm laborer, and the craftsman of an earlier day were like the coal miner of the 1860s and 1870s. All of them gave forth a tremendous expenditure of energy and effort. In short, Wright created a common identity among hard-working, industrious Americans that bridged the gap between skilled and unskilled laborers and even between labor and the professional and business groups.[16]

While Wright defined labor broadly, he showed an awareness that conditions were changing for many who labored. For example, he noted that coal miners worked "in wet, damp and exposed chambers, often a thousand feet below the surface of the earth, . . . liable at any moment to be crushed to death, or maimed for life." Wright claimed that the miner's occupation was "more destructive to health than other occupations above ground." The mines produced "cramps and rheumatism" and lungs became "diseased by the tepid atmosphere of the dark vaults of the subterranean chambers."[17] With the answer a foregone conclusion, Wright asked, "What occupation or employment save war shows such terrible risk of life and limb?"[18] By any understanding of Christian decency, Wright concluded that miners were entitled to better treatment. In a practical sense, Wright noted that the failure to improve wages and conditions would worsen the miners' physical condition, lead to crime and intemperance, increase the number of widows and orphans, and help to entrench a permanently depressed element in American society.[19]

For Wright, the rise of large corporations that exercised monopolistic controls and gained special privileges constituted the root cause of labor's distress and of socio-economic problems generally. Wright pointed to drastic and undesirable changes that had taken place in the coal industry over the prior thirty years. Independent mine owners and operators who had been responsible for developing the coal industry were increasingly forced to confront powerful corporate entities which sought to extort enormous profits from coal-mining operations. Railroad monopolies particularly, through the regulation of shipping rates and even the gradual takeover of the mines, were extracting exorbitant profits from the anthracite fields.[20] As a result, independent operators paid out the bulk of their profits to cover transportation costs, the consuming public paid higher prices for coal, and the miners' wages were frozen in place or actually reduced.[21] From Wright's perspective, big capital, controlling enormous amounts of wealth, was gradually displacing the entreprenuer and inflicting a new set of circumstances upon anthracite miners. It was not, therefore, simply a matter of labor against capital, but rather greedy and unethical capitalists against honest and hard-working businessmen and laborers.[22] Wright could easily extrapolate from his own political background and see that the worst fears of the Jacksonians were coming to pass. Monopolies were gaining undue control over American life.[23]

Wright concluded that the monopolist robbed labor of its rightful profits. Not unlike Karl Marx, Wright argued that laborers rather than capitalists contributed to the basic value of any product. He claimed that capitalists were nothing without labor and that, in fact, the proper relationship between labor and capital should parallel that between a parent and child. He justified "protest against that system of political economy which makes one man do the work" and allows another to "reap the benefit of it."[24] Since workers expended so much energy on the coal that they mined, Wright believed that they were entitled to a relatively fixed "basis" or percentage of the profits. Wright was not prepared to say what that basis should be, but argued that laborers and independent coal operators should reach agreement on it.[25] Although he hoped that such an arrangement would stabilize the price of coal over long periods of time, he saw that the problem with this potentially fair and workable system was the railroad monopoly. Wright knew that the monopolies charged unduly high transportation rates no matter what the price of coal might be.[26] The monopolist, therefore, constituted a villainous force that stood in the way of a proper and lasting arrangement.

Wright also thought that these same monopolists were responsible for creating new and dangerous attitudes toward work and diligence. In a manner that anticipated Thorstein Veblen, Wright noted that the accumulation of enormous fortunes enabled some Americans to devote their lives to entertainment and recreation. For those at the top, the appearance of idleness and leisure was socially more respectable than hard work. The very rich saw manual labor as degrading and, by their life styles, gave evidence that it should be avoided.[27] Wright worried that the well-to-do adopted the phony and decadent values of Europe, especially Paris with its "follies and corruption," and denigrated the hard-work ethic that made America great. What's more, Wright feared that their attitudes penetrated American social values generally, and that many people who could afford to were increasingly prone to imitate those at the top. In fact, Wright claimed that ever larger numbers of Americans were treating what at one time were luxuries as absolute necessities. Wright felt that "the wealthy should furnish a better example," but thought this unlikely "in a country where the only badge of nobility is money."[28]

In response to the monopolists' negative attitude toward the miners and the miserable wages and conditions they inflicted upon them, Wright defended labor's right to organize. Taking issue with corporation lawyers who held that unions constituted illegal "conspiracies," Wright argued that they were perfectly lawful because they intended no harm to anyone. Furthermore, he noted that since business monopolies were in themselves combinations defended by law, legislative bodies had no right to deny laborers similar advantages. Because corporate groups were allowed to fix prices, surely laborers should be allowed to combine together and fix the price at which they would work.[29] Developing the argument still further, Wright defended labor's right to strike. Specifically, Wright argued that the anthracite strike of 1870–1871 was entirely justified. He indicated that railroad and mining corporations which controlled anthracite production in the Wyoming Valley had arbitrarily reduced wages by one-third. The Workingmen's Benevolent Association, a miner's union founded in 1868, could see no alternative but to call a strike. When one surveyed the evidence objectively, Wright wondered how anyone could legitimately doubt the justification for the strike.[30]

But while labor was justified in the use of strikes, Wright doubted the wisdom of such methods. He concluded that neither labor nor the coal operators benefited from strikes. The recent strikes with which Wright was familiar benefited only "those persons who own the shares of the great carrying and trading monopolies." These monopolies had "millions of tons of coal piled up at the market" which, according to Wright, "more than doubled in price during the strike of the coal miners in the summer of 1869."[31] Utilizing an argument put forth by a number of labor analysts, Wright suggested that laborers never really regained the income they lost during strikes and, from a practical standpoint, were better advised to keep working even when wages were reduced.[32] Furthermore, Wright noted that strikes directed at the small independent operators were pointless since those operators were as much victimized by railroad monopolies as laborers. Faced by rising transportation costs that absorbed their profits, independent owners often had little choice but to reduce the miners' pay.

Instead of trade-union activity, Wright proposed various solutions to labor's woes. Economically, he urged laborers to consider the possibility of forming cooperative associations that could engage in business and agricultural activities. Wright actually linked the call for cooperative ventures with his ideas about strikes. In periods when workers would have ordinarily gone out on strike, they should instead continue to work and contribute their wages to a general fund. Although Wright never bothered with administrative details and failed to indicate how individual workers with marginal incomes would support their families, he argued that miners collectively earned an enormous amount of money and that even ten per cent of that sum would yield a tremendous surplus. This surplus would enable workers to buy their own stores, farms, small industries, and even their own coal mines. Laborers would then be able to buy goods and services at modest prices and accumulate profits. Wright asked why the laborer should "continue on year after year enriching manufacturers, middle-men, and retailers, when he has the whole thing within his own control?" Wright concluded that the worker could just as well be his "own banker, manufacturer, merchant, or farmer as any other person or class of person may be."[33]

But even as Wright suggested such economic alternatives, he believed that labor's most effective effort would occur through political action. Since big capitalists controlled state and national legislative bodies, Wright again and again stressed labor's tremendous numerical strength as potentially the best weapon to use against monopolists. Composing nine-tenths of the population, labor constituted an irresistible force that only needed to assert its power at the polls. Through control of the legislative process, labor possessed the keys to constructive and non-violent change. In the narrow sense, Wright indicated that political change would operate to labor's special advantage; it could secure the specific interests of the laboring community. But Wright suggested that labor had a broader duty than simply class loyalty; he defined labor's responsibility in terms of the entire nation and its people. Labor's task was to save American institutions and values from the monopolists; independent businessmen and the consuming public were not strong enough to achieve this goal without help. Laborers must watch with "vigilance the schemes of bad men, who too often do not hesitate in their ambition to weaken if not destroy the political fabric." Wright believed that laborers, "being in the majority," had a duty "to see no political wrong is inflicted on the State."[34]

There was an undeniable paternalism in Wright's attitude toward labor's political role. Whereas Wright held that labor was to capital as a parent to its child, he reversed the analysis when it came to the political process and the state. He now compared laborers to children wronged by their parents, but entitled to protection from government, the so-called parental authority. Even as Wright expressed confidence that "there would be more practical and beneficial laws" if "there were more laboring men in power," he advised laborers against direct participation in politics and "discouraged them from becoming Political adventurers." Political activity was a thankless task that absorbed the energies and finances of the participants. Wright hinted that political life involved dirty practices that struck at the very roots of labor's nobility. It was best left to men who were hardened to it and could afford, by their financial means, to devote full attention to it.[35] The task for laborers was to educate themselves regarding the issues that affected them and to become fully conversant with the solutions that were available to deal with their plight. Once done, they must investigate the various candidates for office and choose those who, by background and principle, would see to labor's specific interest as well as to the national interest.[36] Wright carefully refrained from advocating any particular individual or party as labor's special friend, but the reader could rather easily conclude that Wright himself would be an ideal choice.

The political program that Wright offered covered a range of public issues. He offered concepts and suggestions that were clearly meant to uplift laborers as well as a number of ideas that aimed to reduce the power of monopolies. Generally, Wright argued that his ideas would benefit a broad cross-section of Americans rather than a single class. The proposals varied a good deal in the details that Wright provided. Occasionally he discussed specific changes at considerable length and revealed an elaborate plan about how they might be implemented. Often his ideas were only vaguely stated; sometimes he only hinted at possible change without setting forth a clear recommendation. At times, he proposed solutions that were relevant to national problems while at other times he directed attention specifically to problems in Pennsylvania. On both the state and national scene, Wright indicated that change would come through legislative action and changes in the law.[37]

An immediate legislative goal that Wright espoused for laborers involved a limitation on the daily hours of work. In particular, Wright favored passage of an "eight-hour" law that would reduce the average work day by two hours. He argued that such a reduction was especially necessary for coal miners who desperately needed physical relief from their arduous tasks. Wright thought it "strange that the banker, the broker, the merchant, and the general business man should now be limited by custom to a shorter number of hours" while laborers worked longer and harder than ever before.[38] In fact, Wright not only argued that the worker's hours should be reduced by law, but that his wages ought to be raised. Although Wright hinted that legislative action might secure the latter point at the same time an eight-hour law was enacted, he did not suggest a specific proposal or method by which it could be accomplished.[39]

Wright thought that an even better hope for labor than wage and hour legislation involved a favorable disposition of the public-land issue. Although the benefits of a strong public-land policy would not immediately ease labor woes, in the long run Wright hoped that proper legislative action would provide a permanent advantage. Wright argued that too much western land had already been granted by Congress to railroad monopolies and speculators.[40] Whereas Wright noted that the Homestead Law intended the land to be used by individual settlers and their families, he pointed out that millions upon millions of acres were now hoarded by monopolists. These lands could have provided laborers with millions "of comfortable homes" upon "broad acres." Although two-thirds of the public domain still remained to be disposed of, and Wright believed that the situation was not yet irretrievable, he feared that if current trends continued, "the public domain will disappear in a short time."[41] The only way to protect labor's interest in the land was through the erection of sufficient legislative safeguards.[42]

In regard to monopolies, Wright conveyed a general attitude that legislative bodies representing labor could somehow restrain them. At the very least, such bodies could deny monopolists the special privileges that they had been granted over a period of years. Wright's most specific proposal had to do with tariff policy. Like other Democrats, Wright held that high protection benefitted corporations and encouraged the growth of gigantic monopolies. Tariffs enabled the capitalists to increase profits while the laborer "remained in the same notch," still living in his "rented tenement."[43] Nevertheless, Wright did suggest conditions by which laborers could benefit from tariff protection. If wages were raised proportionately with each tariff increase so that monopolists could not take the bulk of the increased profits, tariff barriers would be acceptable. In a parallel manner, Wright suggested that if laborers were able to form cooperative associations, and thus create their own business ventures, they could also benefit from tariff protection. Otherwise, tariffs ought best to be reduced substantially.[44]

There was a consistency between the appeal Wright made in the *Practical Treatise* and his own personal circumstance that belies the charge of hypocrisy and political expediency. The problems that Wright pointed out and the solutions he suggested indicate that he was greatly concerned about hardening class distinctions and future conflict. In the course of his own lifetime, Wright saw the increasing polarization of society at both the top and the bottom. By

1870, there were massive numbers of laborers who lived and worked in marginal conditions. Each year they became ever more remote from the mainstream of American life. The poor immigrants, typified by the Chinese whom Wright attacked, caused the working classes to sink lower and contributed to the widening social gap. Wright understood that these workers would increasingly take in unionism, strikes, and even violence to protect themselves and improve their condition. And those who were responsible for this sad state, according to Wright, were the monopolists who held lofty positions at the top of the socio-economic scale. Through their political and economic power, they created entrenched poverty and a rigid class structure. As conditions worsened, Wright thought that the danger of violence and revolution became more apparent.[45]

From Wright's perspective, monopolists posed their greatest danger to the middle class rather than to the laboring class. He knew that his proposals required labor's cooperation and support, and consequently he spoke most directly to that class. But the programs Wright hoped that labor would help to implement were contrived mainly for the salvation of the middle class. By family background and professional status, as well as his emphasis on hard work, individual ability, and social mobility, Wright identified with the middle class. The massive wealth of the monopolies and the utter degradation of the working classes now threatened those middle-class values that Wright cherished. Mobility, individual initiative, and hard work counted for little in a society where rigid class divisions existed. Moreover, the imminence of violence and class warfare threatened to consume the entire system that had once allowed middle-class values to flourish. The confrontation between labor and capital, even though it might someday result in a better situation for labor, would surely destroy those aspects of American life that Wright felt constituted its strongest virtue.

Wright also was concerned about the question of social status. As a lawyer and successful businessman, he saw himself as a spokesman for the middle class; he identified with an upper-middle-class leadership that had been the moving force within American life during the early nineteenth century.[46] He had achieved honor and respect within the context of that class and believed he had thrived by living according to its value structure. With the passage of time, however, he saw that his entire situation was beginning to wither away. Wright himself confronted new monopolistic forces in Congress and in Pennsylvania politics. Independent coal operators and successful businessmen, people who Wright thought were much like himself, were nudged from positions of leadership by corporate groups wielding far greater wealth and power. In many instances, the upper-middle-class leaders who had risen to prominence in the Wyoming Valley and Luzerne County now confronted railroad and corporate monopolies that were external to the region. Based in New York

City and elsewhere, these faceless entities created economic policies without concern for or consultation with the traditional leaders in the Wyoming Valley. For many of the men who associated with Wright, there was bitterness, a feeling of frustration, and for some, a feeling that nothing could be done about it.

To counter this situation, Wright tried to create a broad unity among hard-working, diligent people. Appealing to laborers and middle-class Americans on the basis of a common work ethic and insisting that they collectively composed nine-tenths of the population, Wright hoped to create a powerful alliance capable of destroying the monopolists' power. In addition to Wright himself, the leaders in the struggle were of course meant to be upper-middle-class Americans who had traditionally spoken for American society. These professional and business people would operate within the existing system to heal it of its ills and expunge the evil elements that corrupted it. The political system and the law, those things that had meant so much to Wright for so many years, were to be the principal agents of change.

Once the national and state political systems were brought under control, Wright hoped that a predominantly middle-class society could be re-created. In the first instance, the special privileges that allowed the monopolist his power, his distinctiveness from other Americans, would be eliminated. Secondly, the laboring classes would be raised in status, uplifted to the point that they, as small businessmen, merchants, and even farmers, could identify with the mainstream of American life. According to Wright's plan, the broad middle ground of American society, complete with an operative idea of social mobility and individual effort, would once again be allowed to flourish.[47]

The solutions Wright offered were not very practical. Rather than dealing with urban-industrial conditions directly, he proposed to escape from them. Essentially he did not choose to deal with labor as a fixed part of American life, but hoped instead to offer laborers channels of advancement to middle-class status. Even his call for eight-hour legislation was put forth in this context. Although labor's condition would be improved through such legislation, Wright hoped that a reduction in the workday would encourage laborers to pursue additional educational opportunities that would eventually allow them to rise on the social scale. Wright's desire to maintain public lands for actual settlers had a similar purpose. Many labor advocates wanted to preserve free and open land in the West because the threat of settlement by eastern laborers would force employers to maintain high wages. Wright, however, did not particularly link western land with higher pay; he instead hoped that laborers would actually take possession of the land and thereby change their circumstances.[48] Wright, therefore, in pointing laborers toward education, land ownership, and small business activity, looked back to an America of small towns and forty-acre farms.

It also seems clear that Wright misjudged the cause of America's problems. His analysis suggested something very close to a devil theory of American development. According to Wright, the essential political and economic systems were perfectly all right, it was simply that they had fallen into the hands of evil and unscrupulous men. In a manner that paralleled fundamentalist oratory, Wright condemned monopolists as greedy and selfish people who distorted a great system to their own advantage. There was no awareness that monopolists might be products of industrial capitalism and that the political system was simply unprepared to deal with their influence. Not surprisingly, Wright believed that the answer to evil was simply to put honest and good men in positions of power. These honest men would represent the vast majority of hard-working people in the battle against the evil and indolent few. Wright could not see that the "honest" men of his own class might be corrupted too, especially when they confronted the new conditions of post-Civil War America.

But if Wright's analysis and solutions were naive, he was part of a generation which offered impractical proposals. Terence Powderly, leader of the Knights of Labor and several times mayor of Scranton, philosophically opposed strikes and favored western settlement as a basic solution to labor woes. Patrick Ford, editor of New York's *Irish World* and a leading labor spokesman during the 1870s, also favored free land in the West as the best way to solve labor-capital conflict. Even Henry George, with single-tax theories that once seemed so revolutionary, sought to rehabilitate traditional middle-class America and largely ignored the basic difficulties posed by industrialism.[49] All of these men, at one time or another, justified the existence of trade unions, but none was committed to trade union methods. And several decades later, many Progressive leaders believed, like Wright, that the political system was essentially workable. They wanted to make sure that honest representatives of the majority were allowed to watch over the interests of the nation.[50]

In essence, Wright was something of a reactionary in his desire to return to an older America.[51] He was not opposed to capitalism, but rather what some capitalists had become after the Civil War. He was a product of Jacksonian America who favored equality of opportunity, but who opposed equality in the concrete sense. He believed that those best qualified by ability, by what they had made of themselves through hard work, should be responsible for American government and society. In a parallel manner, he was willing to help those on the bottom to improve their condition, but was too much the paternalist to encourage those same people to work through trade-union methods or direct political activity. But though his basic objectives pointed back to an earlier time, Wright's call for an active, positive government to solve problems looked to the future and the Progressive Era. Although the machinations he used to gain office suggest the political huckster, it seems clear that he philosophically valued political participation, not solely as an end in itself, but as a means to accomplish socio-economic purposes.[52]

NOTES

[1]Frank B. Evans, *Pennsylvania Politics, 1872–1877: A Study in Political Leadership* (Harrisburg, 1966); and James A. Kehl, *Boss Rule in the Gilded Age: Matt Quay of Pennsylvania* (Pittsburgh, 1981).
[2]Oscar Jewell Harvey and Ernest Gray Smith, *A History of Wilkes-Barre, Luzerne County, Pennsylvania,* Vol. V (Wilkes-Barre, 1930), 190; and William Wood to Hendrick B. Wright, 6 November, 16 December 1871, and Rhoda A. Evans to Wright, January 1872 (Hendrick B. Wright Papers, Wyoming Historical and Geological Society, Wilkes-Barre, Pennsylvania).
[3]Dumas Malone (editor), *Dictionary of American Biography,* XX (New York, 1936), 554. For similar estimates of Wright, see the New York *Times,* 28 July 1877; and Wilkes-Barre *Record of the Times,* 10 April 1878.
[4]George R. Bedford, "Some Early Recollections," *Proceedings and Collections of the Wyoming Historical and Geological Society,* XVI (1918), 59–61.
[5]For information on Wright's boyhood and career prior to the Civil War, see Daniel J. Curran, "Hendrick B. Wright: A study in Leadership" (Unpublished doctoral dissertation, Fordham University, 1962); Hendrick B. Wright, *Historical Sketches of Plymouth, Luzerne Co., Penna.* (Philadelphia, 1873), 402–419; Harvey-Smith, *History of Wilkes-Barre,* 25, 190; and Malone, *Dictionary of American Biography,* 553–554.
[6]Evans, *Pennsylvania Politics,* 5, 18. For information about the early development of anthracite mining, see H. Benjamin Powell, *Philadelphia's First Fuel Crisis: Jacob Cist and the Developing Market for Pennsylvania Anthracite* (University Park, Pennsylvania, 1978).
[7]Evans, *Pennsylvania Politics,* 20.
[8]Malone, *Dictionary of American Biography,* 554.
[9]Hendrick B. Wright, *A Practical Treatise on Labor* (New York, 1871), 19.
[10]Wright, *Practical Treatise,* 46–47.
[11]Wright, *Practical Treatise,* 24–25, 28, 50–51.
[12]Wright, *Practical Treatise,* 107–132. For negative comments about the Irish in the coal regions, see the Pottsville *Miners' Journal,* 8 January, 7 May, 28 May 1870; and the Wilkes-Barre *Record of the Times,* 22 March, 19 July, 2 August, 20 September 1871. Wright, on the other hand, won the favorable attention of Irish-Americans as early as the 1840s by helping to organize efforts to relieve the famine in Ireland. See the Wilkes-Barre *Times Leader-Evening News,* 15 March 1967.
[13]*Irish World,* 17 September 1881. The Hendrick B. Wright collection offers some indication of the attention that Wright and the *Practical Treatise* drew. See specifically James McCosh to Wright, 23 January 1872, S. G. Morrison to Wright, 20 April 1872, John Wallis to Wright, 29 April 1882 (Wright papers, Wyoming Historical and Geological Society).
[14]*Irish World,* 17 September 1881; and Evans, *Pennsylvania Politics,* 137, 178–179, 180–192.
[15]Wright, *Practical Treatise,* 21 215.
[16]Wright's identification of all hard-working people within the same category is also suggested in the history of Plymouth that he published as a book in 1873. He noted

that hard work bound different occupational groups together. See Wright, *Historical Sketches of Plymouth,* 273, 276–277.

It is also notable that Wright's view of personal effort and diligence was strikingly similar to older Jacksonian concepts. See Marvin Meyers, *The Jacksonian Persuasion: Politics and Belief* (New York, 1960), 18–24.

[17] Wright, *Practical Treatise,* 294–296.

[18] Wright, *Practical Treatise,* 303.

[19] Wright, *Practical Treatise,* 92, 213–215, 306.

[20] Wright, *Practical Treatise,* 145–146, 149–150. For references to others who feared growing monopolistic influences, see the Wilkes-Barre *Record of the Times,* 28 November 1868, 13 November 1869.

[21] Wright, *Practical Treatise,* 59–60, 140–142, 391, 396.

[22] During the period in which Wright wrote, monopolistic trends were becoming increasingly evident in the anthracite region. For example, Franklin B. Gowen, president of the Reading Railroad, was successful in developing a monopoly of transportation facilities in the southern fields. Between 1871 and 1874, Gowen also bought large areas of coal lands for the railroad. This economic power enabled Gowen to exercise tremendous influence over both coal operators and the miners. See Clifton K. Yearley, Jr., *Enterprise and Anthracite: Economics and Democracy in Schuylkill County 1820–1875* (Baltimore, 1961), 183–185; and Anthony Bimba, *The Molly Maguires* (New York, 1932), 42–43, 56–57.

[23] Wright represented a part of the Jacksonian movement which had supported government aid to new businesses and internal improvements. In the 1830s and 1840s he supported a protective tariff and aid to railroads. Although these ideas left him personally at odds with Jackson's policies. Wright's emphasis on equality of opportunity at least allowed him to identify with the broad mainstream of the Democracy. For other Jacksonians as well, the desire for social mobility and individual opportunity was the key criteria in determining their participation in Jackson's party. Because of this, they were willing to overlook Jacksonian opposition to positive government. Later, when Wright found that government aid had produced monopolies, he changed his mind about the propriety of such aid. At that point, Jacksonian fears about the development of monopolies and an aristocracy of wealth made more sense to Wright. For a discussion of various individuals and groups who found a home in Jackson's camp, see Meyers, *Jacksonian Persuasion.*

[24] Wright, *Practical Treatise,* 225.

[25] Wright argued that when coal prices did rise, labor's share of the profit should also rise. In a like manner, labor would receive less if coal prices fell. See Wright, *Practical Treatise,* 134, 137.

[26] Wright, *Practical Treatise,* 142, 144.

[27] Wright, *Practical Treatise,* 219–220, 262, 296; and Thorstein Veblen, *The Theory of the Leisure Class* (New York, 1967).

[28] Wright, *Practical Treatise,* 217–219.

[29] Wright, *Practical Treatise,* 74–75, 78–79.

[30] Wright, *Practical Treatise,* 182–183, 383–385, 391.

[31] Wright, *Practical Treatise,* 183.

[32] Wright, *Practical Treatise,* 184–185; and Terence V. Powderly, "The Organization of Labor," *North American Review,* 135 (August 1882) 118–126.

[33] Wright, *Practical Treatise,* 184–187.

[34] Wright, *Practical Treatise,* 36–37, 146.

[35] Wright, *Practical Treatise,* xv, 37–39, 264. Terence Powderly of the Knights of

Labor held similar views about labor's participation in politics. He believed that politics would only corrupt and "defile" laborers. See Terence V. Powderly, *Thirty Years of Labor, 1859 to 1889* (Philadelphia, 1890), 44.

[36] Wright, *Practical Treatise,* 39, 264.

[37] Throughout the *Practical Treatise* Wright stressed the idea that legislative action would bring desirable change. Perhaps because of his personal background as a lawyer and as a politician in the state and national legislatures, Wright left little room for executive action.

[38] Wright, *Practical Treaties,* 216.

[39] Wright, *Practical Treatise,* 206–231.

[40] Wright, *Practical Treatise,* 244. During the 1870s, Wright made an active effort to push legislation through Congress that would regulate the disposal and use of public land. See the *Irish World,* 3 January, 10 January, 10 April, 24 April 1880.

[41] Wright, *Practical Treatise,* 330, 333.

[42] Wright, *Practical Treatise,* 327–344. In later years, Wright took an active interest in currency inflation as a concept beneficial to laborers. This is indicated by his role in the Greenback Labor party. However, the *Practical Treatise* discussed currency matters in only the most peripheral ways. See Wright, *Practical Treatise,* 228–229. For a discussion of Wright's changing views of inflation, see Curran, "Hendrick B. Wright," 224, 278–279, 295.

[43] Wright, *Practical Treatise,* 66, 270.

[44] Wright, *Practical Treatise,* 284–287. Wright also hoped that changes in the National Banking System would help to minimize the monopolists' influence. He argued that big capitalists, with government help, controlled that system for their own benefit. Furthermore, Wright argued that tax exemptions on the interest that government bond holders earned should be eliminated. Wright believed that only the very rich held such debt issues and the poor were thereby being discriminated against. See Wright, *Practical Treatise,* 310–311, 318–326.

[45] Wright, *Practical Treatise,* xiv, 35, 40–41, 381–382.

[46] In his historical discussion of Plymouth, Wright demonstrated his high regard for middle-class spokesmen and leaders. He described at length how physicians, merchants, schoolmasters, and ministers had been responsible for the town's development. On the other hand, people with large amounts of money were not apparent in contributing to that growth. See Wright, *Historical Sketches of Plymouth,* 279–284, 303–312, 332–350.

The threat to status and power that worried Wright and others shortly after the Civil War suggests similarities to the problems that bothered middle-class Progressives in the early twentieth century. See Richard Hofstadter, *The Age of Reform, From Bryan to F. D. R.* (New York, 1955), 131–173.

[47] Wright's reverence for a broad consensus within American society is suggested in his historical work on Plymouth. For example, he noted that during "harvest time the minister, the schoolmaster, the blacksmith, the wheelwright and the carpenter lent a hand, and all went 'merry as a marriage bell.' " Moreover, Wright claimed that "the tradesman, the merchant, the mechanic, and the farmer, as to social caste, all stood upon the same platform." Wright believed that this was a desirable social situation. See Wright, *Historical Sketches of Plymouth,* 273, 277.

[48] For a discussion of the issues involved in western settlement and their relevance to labor, see Helene S. Zahler, *Eastern Workingmen and National Land Policy, 1829–1862* (New York, 1941). Wright's call for western settlement was especially inappropriate for Irish-American laborers. Most Irish immigrants were disgusted with farming

because of their experience in Ireland. They also feared the loneliness of the American West and lacked the capital to participate in commercial agriculture. See Wayne G. Broehl, *The Molly Maguires* (Cambridge, Massachusetts, 1964), 83.

In contrast to Wright, others in the coal region thought that western land did not offer Irish laborers much hope and instead urged Irish-Americans to help each other to find jobs. See the Pottsville *Emerald Vindicator,* April, December 1875, March 1876.

[49]Daniel Aaron, *Men of Good Hope: A Story of American Progressives* (New York, 1961), 55–91; Powderly, *Thirty Years of Labor,* chapter V; *Irish World,* 12 July, 2 August, 23 August 1879.

[50]Hofstadter, *Age of Reform,* 203–204, 267–268.

[51]A comparison of Wright's two published works indicates the author's high regard for an earlier style of life. In the *Historical Sketches of Plymouth,* Wright often romanticized the conditions of rural life in pre-Civil War America. His *Practical Treatise* expressed regret that the old conditions were gone. See Curran, "Hendrick B. Wright," 271.

Based upon my own reading of the two works, I found that the *Practical Treatise* offered suggestions about how to restore the older society along with its value structures.

[52]It is interesting to speculate upon Wright's influence. The impractical nature of his proposals, combined with his largely regional influence and reputation as a demagogue, meant that he would not seriously influence the mainstream of American reform. Nevertheless, it is possible that he had a profound influence upon conditions in the anthracite region. Victor Greene has argued that the new immigrants of the 1890s were the first to develop an effective union movement and that older immigrant groups, by implication the Irish, were ineffective in their organizational efforts. Indeed, it seems clear that Irish institutional development in the anthracite region never paralleled the effective religious, political, and economic organizations they developed in New York and Philadelphia. In part, the expansive anthracite geography, the lack of a majority of the population, and the nearness to Irish-American spokesmen in New York City and other places may explain Irish organizational weakness in the anthracite region. But it also is possible that men like Wright, as well as Powderly, who both were well regarded by the Irish, may have blunted the development of a stronger response because their theories were popular. In this sense, Wright, instead of achieving consensus, may have helped pave the way for greater conflict. By postponing consideration of realistic solutions, he helped insure the development of an even more explosive situation. See James P. Rodechko, "Irish American Society in the Pennsylvania Anthracite Region, 1870–1880," in John E. Bodnar (ed.), *The Ethnic Experience in Pennsylvania* (Lewisburg, 1973), 19–38; and Victor Greene, *The Slavic Community on Strike* (Notre Dame, 1968).

Chapter Six

COMMENTARY: CORPORATE ATTITUDES TOWARD LABOR ORGANIZATIONS AND HENDRICK B. WRIGHT

Melvyn Dubofsky

James P. Rodechko's and Perry Blatz's well-crafted papers have a remarkably contemporary ring, just as they also demonstrate the virtues of solid research in local and regional history. Mr. Rodechko's paper, especially, reminds me of the recent New York gubernatorial primary campaign in which the victorious Republican candidate, a successful small entrepreneur or, at least, the heir to the fortune from a family drugstore chain, stressed his sympathy for working people and his ability to create jobs. In New York we heard and saw on television more than seven million dollars of the candidate's money devoted to the jobs theme, and over the next few weeks we will be inundated with a few million dollars more of Lewis Lehrman's television advertising. Lehrman's 1982 campaign to become governor of New York suggest precisely the questions raised by Mr. Rodechko's paper: how far are politicians motivated by crude opportunism, how far by solid principle?

Mr. Blatz deals with the timeless issue of labor-capital, worker-management relations. His paper, too, suggests current headlines about trade-union defensiveness and give-back bargaining. And he raises the question of how far employers, even the rhetorically most paternalistic and considerate among them, really care for or serve the interests of their employees.

In his well-wrought analysis of the politics and principles of Hendrick Wright, Mr. Rodechko deals with one of the thorniest problems historians face, the concept of "false consciousness": to what extent people are really aware of their true interests and to what extent do they deceive themselves. In the case of Hendrick Wright, was he no more than an opportunistic demagogue or did he sincerely believe in what he said and did?

As I read Rodechko's paper, Wright appeared a man with no conscious or even unconscious conflict in his approach to voters. Wright truly believed what he said, and what he said, to be sure, served his opportunistic purposes. For reasons I will now enumerate, Wright was very much a man of his age and place. Like most Americans of his time, he was caught up in what might be called the "producer ethic" and the "Lincoln myth." With Abraham Lincoln, Wright concurred that "the laborer of today is the capitalist of tomorrow." Like Lincoln, Wright could place human rights over property rights, the laborer over the hirer, yet never seek to do anything that would threaten private property or capitalism as a system. Also, like Lincoln, Wright rose from humble agrarian beginnings, first to success and wealth as a corporation attorney and then to party politics and election to the House of Representatives. The Pennsylvanian, Mr. Rodechko tells us, also harbored presidential ambitions.

Wright's anti-corporate, anti-monopoly themes also resonated with popular culture and sentiment in the mid- and late nineteenth century. Wright's rhetoric offered a perfect example of historian Daniel Rodgers' point about the crisis the "work ethic" faced in late nineteenth-century America.[1] How could the imperative to earn bread "by the sweat of the brow" remain defensible when set against the reality of intrinsically unsatisfying industrial detail work and the conspicuous consumption of the "robber barons"?

Wright's spirited defense of unions and strikes calls to mind Herbert Gutman's analysis of industrial conflict in the late nineteenth-century United States. In Gutman's model, local entrepreneurs and professionals tended to ally with the community's workers to fight national, or alien, capitalists.[2] In sociological jargon, Wright might be described as a leader of the party of the provincials battling the power-grabbing cosmopolitans.

Wright's stress on producer cooperation and political action in preference to such direct labor action as strikes reminds one of the policies and practices of the Knights of Labor. Wright and Terence Powderly, the Knights' Grand Master Workman and twice Greenback-Labor mayor of Scranton, were two peas in a common Gilded Age pod.

Finally, it must be remarked that in Hendrick Wright's worldview, no conflict existed between the interests and needs of the middle class and the working class. The only conflict that Wright perceived was the one between producers and parasites, those who created real values versus those who dealt in paper. To repeat myself, Wright was a man of his time and also an American for all times. Like most of his predecessors, contemporaries, and successors, he never resolved the contradiction between the principles of equality of opportunity and equality of condition. Throughout his life Wright bewailed the enormous inequality of condition in the United States that resulted precisely from the operation of the equality of opportunity that he sanctified.

I have just two minor quibbles with Mr. Rodechko. First, it would have been helpful to see in some detail how Wright's political practices and achievements, if any, compared with his rhetorical principles. Second, I wonder why Pennsylvania Democrats were so slow to appeal to Irish Catholic immigrant miners.

Recent scholarship concerning ethno-religious factors in American political history indicates that the national Democratic party had begun to build a firm political constituency among Irish Catholic voters long before the Civil War.³ Why, then, was northeastern Pennsylvania different?

Turning now to Mr. Blatz, we observe how his paper strips away the veil of "paternalism" from the professed concern of capitalists for the welfare of their employees. He also shows quite clearly that the core issue involved in labor-management relations in 1888-1889 was what it had been earlier and would remain later, *power*. Wages, hours, and conditions could always be negotiated or compromised, but for capitalists, the distribution of power could not. It was easier to make concessions to non-union than to union workers. In the case of the former, charity, benevolence, or human concern were operative; in the case of the latter, the protection of capital's authority and its right to control labor prevailed. Employers dealt with unions only under the lash of necessity. That remains the same today as it was in 1888-1889.

Blatz's paper, like Rodechko's, also touches the issues of "false consciousness" and "opportunism." Did any capitalist really believe in paternalism, whether of the variety described by Eugene Genovese or Richard Sennett, as Wright quite obviously believed in the "producer ethic" and the "Lincoln myth"? Blatz's employers, their paternalistic language notwithstanding, treated their workers as bosses, not as doting fathers or kind uncles.

It was indeed a pleasure to read two such fine papers as those prepared by Messrs. Rodechko and Blatz. I hope that my remarks encourage them to pursue their research further and to tell us more about the history of northeastern Pennsylvania and the people who lived and worked there.

NOTES

¹Daniel T. Rodgers, *The Work Ethic in Industrial America, 1850-1920* (Chicago, 1978).
²Herbert G. Gutman, "The Worker's Search for Power: Labor in the Gilded Age," in H. Wayne Morgan, ed., *The Gilded Age: A Reappraisal* (Syracuse, 1963), 38-68.
³See for example, Lee Benson, *The Concept of Jacksonian Democracy: New York as a Test Case* (Princeton, 1961): Richard Jensen, *The Winning of the Midwest* (Chicago, 1971); Paul Kleppner, *The Cross of Culture: A Social Analysis of Midwestern Politics, 1850-1900* (New York, 1970) Samuel McSeveney, *The Politics of Depression: Political Behavior in the Northeast, 1893-1896* (New York, 1972).

Chapter Seven

THE FAMILY ECONOMY AND LABOR PROTEST IN INDUSTRIAL AMERICA
HARD COAL MINERS IN THE 1930s

John Bodnar

Numerous historians have maintained that the pervasive labor protest of the 1930s in America did not go far enough in effecting social and political change. Disappointed that industrial workers seldom went beyond issues of job security, control of production, or establishing bureaucratic unions, these scholars have insisted that labor possessed an inherent militancy which was repressed by external forces. David Montgomery, a leading interpreter of American labor, has argued that the New Deal, for instance, was simultaneously liberating and co-optive since it eased the burden of managerial control but subjected powerful unions to tight legal and political control. Other analysts of the labor upsurge have blamed the federal government for creating a "businesslike" relationship between labor leaders and management which tempered the "explosive creation" of the mid-thirties. Even the recent surge of interest in the workplace as a framework for discovering the nature of working-class behavior continues the theme of workers as victims and rests on a questionable assumption that working-class militancy was rooted solely in the structure of the factory. Nelson Lichtenstein has argued this point recently and observed that the preoccupation with workplace objectives inhibited the realization of larger social and political goals. Montgomery has echoed this explanation of worker protest and continued to locate the sources of working-class "conservatism" outside the world of labor by stating that their attempts to control the process of production was destroyed by managerial offenses, such as scientific management which attempted to enforce industrial discipline.[1]

A considerably different view of working-class protest emerges if analysis is carried beyond the traditional foci of workplace and government to the familial and cultural systems of industrial workers. While recent years have witnessed an increased interest in this aspect of labor history, few attempts have been made to link specific values and goals discovered among workers to the protest they frequently exhibited. But an opportunity to view the association between the culture of industrial workers and their politics exists with anthracite coal miners in the two decades before 1940. Clustered in the hard-coal region of eastern Pennsylvania, the men and women of anthracite faced difficult circumstances during the period as their industry slowly lost ground to alternative fuels such as gas and oil; the supply of jobs was steadily declining. Faced with the growing spectre of economic collapse, these workers as much as any in the 1930s were forced to identify what in life they valued most and defend it.

Two objectives emerged from the ranks of anthracite mining people in reaction to severe unemployment and depressed conditions after 1930. Central to their goals was a fervent desire to share available work opportunities, which they usually called "job equalization." Equally as important was a demand that the United Mine Workers of America be eliminated as their representative because of widespread corruption which enabled union leaders and coal operators to thrive at the expense of the rank and file. Indeed, this corrupt relationship was thought to be a major impediment to securing the ultimate panacea of job equalization, since corrupt union officials favored some men over others in distributing work.

An understanding of the tremendous rank and file sentiment for these objectives is important because they were not the only alternatives mining people in this region had available to them during hard times. Like all industrial workers they did have concerns about exerting control at the workplace and, as described below, were frequently able to do so. Various labor organizers visited the region and attempted to organize men into "unemployed councils." Local politicians, fearing that these councils offered fertile ground for militancy, countered with the organization of "unemployed leagues." Community and religious leaders continually called for restraint. But despite urgings from both the left and the right, most miners in the northern fields went forward with their eyes riveted to union reform and job sharing; working-class protest, as one radical organizer admitted, was a creation of the region's own workers.[2]

Job equalization became the central objective of anthracite people in response to intensifying hard times after 1930. In brief, equalization meant the even distribution of the mining and processing of hard coal among the various collieries owned by a company, rather than their restriction to the one or two in which the coal may be closer to the surface and, therefore, cheaper to obtain. To an extent, miners were challenging the company's right to control the production process and to make management decisions solely on the basis of cost efficiency.

Throughout the hard-coal region, sporadic protest for job-sharing emerged as unemployment climbed after 1930. In the Panther Valley area, an equalization committee was formed by miners and local business and civic leaders to press the issue with the Lehigh Coal and Navigation Company. In August 1933, fifteen thousand men went on strike in the valley for "equalization of work time." The following year the Lithuanian Catholic Action Convention in Girardsville adopted a resolution urging Franklin D. Roosevelt to compel operators to equalize work "even if no profits are realized and executive salaries are reduced." Similar demands came from the unemployed at the Loree Division of the Hudson Coal Company and from miners in Kingston and Shenandoah.[3]

Naturally, coal operators opposed such plans. They insisted that they be allowed to "concentrate" production at a few collieries where labor and production costs would be least. Such practices favored, especially, operations where coal did not have to be hoisted very far or was obtained with a minimum of blasting and rock removal. In fact, such practices had created unemployment levels so high by 1933 in communities such as Tamaqua, Coaldale, Nanticoke, and Shenandoah that employment had dwindled to one-third of the 1929 levels. Despite recommendations by Secretary of Labor Francis Perkins for equalization, the operators argued strongly against the plan and claimed that it would erode further the already weak competitive position of anthracite coal relative to newly emerging fuels such as oil and gas.[4]

At some local collieries, the men did not wait for political dialogue to settle the matter, but instituted plans of their own to allow greater numbers access to employment. At Glen Lyon, for instance, miners were able to overcome company opposition and establish a limit on the number of cars each man could load. If a miner was stationed in an area where coal was easily accessible, he would have a lower limit than men who could fill their cars only with a great deal of difficulty. As one miner recalled, "In most places it was five cars a day, but on some gangways it was four. If you were over the limit you would pay a fine." The goal of all this was, as the men themselves stated, to insure that each man got a fair day's wage and to "make the mines last longer for everyone."[5]

The entire issue of sharing the work came to a pivotal point in 1933 when the Roosevelt administration conducted hearings to establish National Recovery Administration (NRA) codes for the anthracite industry. John L. Lewis and the UMW had never been enthusiastic over equalization and were concerned primarily with higher wages and shorter hours. Indeed, one scholar has found that the UMW hierarchy was hopeful that Lewis' association with Roosevelt and the resulting prestige he would gain would mute the clamor for equalization, and hard-coal miners would put their faith in the union leader.[6] While Thomas Kennedy, UMW national secretary, did call for equalization, it was a lukewarm endorsement made only after impassioned pleas by anthracite-region people who traveled to Washington to press their demands at the

NRA hearings. Representatives of the insurgent miners organization, business associations, and local equalization committees "all argued for a wider distribution of the available work so that miners can "maintain themselves, their families and their homes."[7]

The fragmentation between operators opposing equalization, union officials straddling the issue, and a working population demanding it resulted in a failure to achieve a consensus necessary for establishing an NRA code in the industry. While such codes would probably not have solved the economic problems of anthracite, the failure contributed to a further erosion of rank and file support for the UMW in the area and intensified sympathy for a "dual union" which emerged in a wave of violence in 1933.[8]

Contemporaries who witnessed the bombings, beatings, and violence ravaging the northern anthracite fields between 1933 and 1936[9] tended to reduce the issue to one of competing union groups fighting for power. But the motivations of the new United Anthracite Miners of Pennsylvania (UAM) ran deeper than that. While some UAM leaders may have seen the UMW as simply a rival for influence, in the eyes of its rank and file supporters the old union of John L. Lewis had grown corrupt. While the anthracite industry may have suffered from the larger economic issue of cheaper, competitive fuels, the rank and file formulated the issue in more immediate terms: the UMW was self-serving and an obstacle to stable employment opportunities.

Interviews with men who participated in these events uncovered the low esteem in which the UMW was held. Chester Brazina, who was head of a UAM local at the Avondale Colliery, had asked for UMW help in 1934 when so many men were getting two days work a week. While no financial assistance was granted, the UMW did raise funds to repair the home of one of its officials, John Boylan, which had been dynamited by rival unionists. "This really threw everyone in an uproar," Brazina remembered. "Everyone started throwing chairs around and one said the hell with them [UMW]."[10]

Dissatisfaction with the UMW was spreading. Joseph Krievak recalled that at Glen Lyon the union officials and the bosses worked together. It was the union officers who usually got "better jobs" where coal was easier to mine. John Sarnoski, an immigrant miner from Poland, was furious that UMW officers would leave work for two weeks to attend conventions and still get paid for loaded cars. Such arrangements tempered considerably the protestation of local UMW officials when they represented the men at their collieries who were blatantly cheated out of their pay when companies counted five loaded coal cars as four. Bart Sheehan, a private detective hired by the coal companies to investigate graft among mining bosses, told a federal investigation in 1933 that bosses regularly sold jobs and put men sympathetic to the UAM in locations where coal was more difficult to mine. And officials of large employers, such as Glen Alden Coal Company, had good reason to strengthen ties with the conservative UMW because they feared the more radical National Miners Union, which began speaking to their employees in 1933.[11]

The situation was clarified by a miner at Wanamie. Leonard Wydotis explained that he supported the UAM because of the "collusion" between the UMW and the operators and felt each miner should get a good day's pay and an "equal amount of work." Wydotis explained that it was difficult, however, to secure fair treatment and equalization if "men on the grievance committee were given the best jobs by the companies." Wydotis explained further:

> The company and the union leaders went hand in hand. So if you didn't support them (UMW) you wouldn't have no jobs. Hell, the boss would go around and tell you who to vote for when we elected union officers. If you insulted the president of the local you wouldn't get a job no place, not even at another colliery.[12]

Miners not only perceived unscrupulous practices on the part of UMW leaders, but also had definite ideas as to the origin of the problem. Invariably, men attributed the growing separation between union and rank and file interest to the initiation of the check-off system. In its 1930 contract, the UMW had secured a "check-off" system whereby the mens' union dues were paid from their wages directly to the unions by the company before the workers ever saw their checks. The complaints of Thomas Maloney, UAM President that their dues were being used for "pianos, pigs, and other articles" and that the system should be abolished angered UMW officials, whose union structure depended on this regular economic infusion.[13] Even more perceptive was the assessment of a Nanticoke miner. George Tresniowski felt strongly that the UMW began to drift apart from the ranks soon after 1930, when it no longer had to deal directly with the men to obtain dues payments. Tresniowski reasoned:

> Before, you had to go to the union hall to pay your dues and then you would see the other members. If there were any problems you would thrash it out right there. After the check-off system the union got the money, but the people didn't go to the meetings; and the union was doing whatever they wanted with the bosses. . . . Before, if you had a grievance you would holler at the union officials. After the check off you hardly saw them.[14]

Since UMW officials were supported directly by payments from companies through the check-off, they could only support attempts by the companies to destroy the UAM. In a remarkably candid recollection, William Everett, a superintendent for the Glen Alden Coal Company in the 1930s, indicated just how effective company recriminations could be. In addition to denying UAM supporters jobs, Everett explained how he dealt with Maloney supporters at his collieries. He recognized that most of the men were for the UAM, so he designed a plan to weaken certain UAM "radical" leaders. One foreman in particular, who spoke loudly for the insurgent union, was sent by Everett to another operation sixteen miles distant so that his journey to work would be very difficult. As he left, Everett warned him, "If I ever hear why you were transferred from any source, you gonna be out of a job." At another Colliery he went to the home of a UAM leader, Mike Savilich, and told him to "tone down" his UAM activities. When the man's wife screamed at Everett, "He's not going to do a thing for you," Everett decided to punish him in a subtle manner. The next day he placed a totally incompetent worker in Savilich's work gang in order to curtail his production capacity. Everett proclaimed, "You had to handle the people, manipulate them."[15]

Everett's attack on UAM leaders could often take the form of granting favors. At the Truesdale Colliery near Wilkes-Barre, he approached the UAM official at that site, Victor Matusa. Everett asked Matusa what made him a "radical." The miner responded that for years his father toiled at Truesdale at a "Catholic place"—a place where coal was so hard to remove it had to be carried as Christ carried a cross. The next day Matusa's father was moved to a better mining station and shortly thereafter Everett secured medical assistance for Matusa's son, who had a deformity.[16]

While the assault of the UAM upon the old union was necessary to obtain its larger goals of widespread and stable employment, a central question remains as to why the rank and file agitated for a sharing of work opportunities. Working-class leaders in the UMW had emphasized wages and hours, and certainly radical organizers were active in the area. It is true that UMW leaders had been discredited, but explanations of worker goals run deeper than simply a reaction to those they may have disliked. A clue to unraveling the web of rank and file motivation can be glimpsed by looking briefly at the manner in which the insurgent unionists conducted their protest.

The salient feature of the UAM protest was its collective nature. Supporters of Thomas Maloney and the UAM attempted to curtail operations at various collieries by stopping work themselves and prohibiting UMW sympathizers from getting to the collieries altogether, or by intimidating them through dynamiting. Invariably this was done in a communal way, with not only the men but women and children in mining families participating.

Women were present everywhere in UAM activities. Several thousand joined the women's auxilliary of the UAM. Led by such spokeswomen as Mary Bernatovich, the auxiliary secured the support of crucial unemployment councils in the area for UAM activities, and recruited speakers who could make appeals not only in English but in Polish and Lithuanian as well. Not surprisingly, when UAM leaders like Maloney were arrested in 1935 and held in contempt of court for violating a strike injunction, twelve of the ninety defendants were women.[17]

At the front ranks, women stood beside their husbands, fathers, and brothers in an attempt to stop UMW loyalists from going to work. During the 1935 violence, women and children stoned men who were attempting to go to work in the Hanover section of Nanticoke. In Plymouth, five women were arrested by State Police for beating a miner on his way to work. Interestingly, the local press withheld the name of the victim but printed the names of the women. In fact, the arrest of the five prompted an angry charge from the UAM women's auxiliary that State Police trying to maintain order in the region had abused women and that "women should cooperate with the new union." In Wilkes-Barre, shortly afterwards, three miners were beaten by ten women who wrestled them to the ground. At Wanamie one evening, women were found "defying" State Police who had been sent to the region to restore order. At high schools in Nanticoke and Wilkes-Barre, students organized support activities for the UAM. Over three-hundred students called a strike at Nanticoke because several teachers had relatives working at Glen Alden Collieries.[18] Communal protest reached a peak when mining families stoned the funeral procession of a miner who was killed while working during the 1935 strike, and filled the victim's open grave with tin cans. George Tresniowski remembered that police had to be called the next day to prevent the protestors from removing the body from the grave.[19]

It was no accident that the UAM protest, emerging as it did from the ranks of the mining communities, involved entire families and communities. Neither was it accidental that this protest ultimately sought a sharing of available work opportunities. For sharing and cooperation were central to the culture these mining people had established by the third decade of the twentieth century. And nowhere were these concepts better learned and employed than in the daily operation of family life. If the culture and lives of mining people had revolved around cooperative familial and communal endeavors, it was reasonable to assume that, given the chance to express themselves, these people would attempt to apply the same concepts to solving larger economic problems as well. Whether solutions were viable to the sagging anthracite industry is not the question. They probably were not. But they were felt to be viable by mining families and were pursued despite retribution by company and union officials. The Glen Alden Coal Company, for instance, initiated a systematic campaign to evict UAM sympathizers out of homes they rented from the company. This move resulted in a stream of protest letters to federal officials such as Frances Perkins, the Secretary of Labor, and Franklin D. Roosevelt. Humble people sat down and wrote about a need for justice, the poverty they endured, and asked Roosevelt for help to "save my family."[20]

An understanding of this family-based culture and economy can be seen in the recollections of children who matured in such an environment. The general pattern of their lives was clear. Childhood was followed by an early departure from school and an entry into the workplace as soon as possible. Earnings were turned regularly and routinely over to parents and this support was rendered at least until marriage, but often throughout the course of entire lifetimes.

Children moved into wage-earning jobs at a feverish pace. Antoinette Watchilla was raised in a family of five children. Characteristically, by age twelve she entered a silk mill in Nanticoke and brought her younger sisters into the mill as they reached adolescence. Her brother had already left school at age eleven to work for a farmer near Kingston. Viola Mazza hired out for house work at age twelve while her sister went to a pants factory to sew. Marie Skovranski was born in 1914 and never left home. Since her mother was a dressmaker, Marie simply stayed home and assisted her. By the time she was fifteen, in 1916, Sophia Reakes had worked as a dressmaker, in a cigar mill, and as a cook in a mental hospital. Adelle Winarski left school following the sixth grade to seek employment in a silk mill.[21] Many young girls who did manage to finish high school left the coal region immediately afterward to obtain housework for wealthier families in the New York City area. Irene Wieczorek claimed that if you were Polish you had such a good reputation as a houseworker that employment came easily. Stacia Tresniowski, who temporarily left Nanticoke for New York in the early 1930s, was able to send sixty per cent of her twenty-five dollar monthly income back to her parents.[22]

Young men, of course, repeated the pattern of their sisters, except that they worked first around the mines in breakers or as "nippers" opening doors for loaded coal cars and spraging them when they had to be stopped. Underground work, usually as a laborer with a contract miner, did not begin much before the age of eighteen. Typical was the case of Steve Stasher, who entered the employ of the mines in Glen Lyon at age sixteen, and helped support his family of origin for seven years until his younger brother was old enough to enter the mines and his sister could go to New York to find work as a housekeeper.[23]

The fact that familial obligations directed wage sharing and entry into work life in the first place was reinforced by the dependence on kin for access to available jobs. This process was crucial for those from abroad. For instance, Lillian Niziolek's father came from Poland to the Susquehanna Coal Company because a brother already in Nanticoke secured an opening for him. Lillian's father, in turn, brought six cousins and nephews to the same mines. Not infrequently, sons would labor with their fathers, as Stanley Salva did in Glen Lyon, or women would talk to a "floor lady" about work for their sisters in silk mills. When work was scarce, particularly during strike periods, workers would move to other regions to live with kin and earn wages, usually returning when work resumed.[24]

While it is clear that families and kin assisted each other, it should not be assumed that this was done in a perfunctory manner. While family tensions existed and obligations generated inevitable pressures, family members—especially children—felt strong obligations to assist their kin, and frequently did so with a deep sense of purpose. The basis for this attitude seemed to be a widespread assumption on the part of adolescents that their parents had worked hard and struggled all their lives and deserved all the help progeny could render. Stanley Salva explained that it was expected that as soon as one was old enough and strong enough, he went into the mines and earned wages. He emphasized, "That's what I wanted." Helen Klish left school after grade eleven and began working in a Nanticoke restaurant because she saw her mother worrying about the mortgage on the family home and felt she was "old enough to owe her that much."[25] As a young girl of twelve in 1918, Mary Yukenevich toiled in a tailor shop "because I wanted to help my mother and father." Antoinette Watchilla revealed that she could not concentrate on going to school "when we needed money in the house." Because she admired her mother for working hard and remaining uncomplaining, she felt compelled to earn wages and turn them over. Virginia Vida and her brothers helped their parents in Nanticoke because they "felt sorry" for them for laboring so hard.[26] And Helen Hosey crystalized the feeling when she explained:

> You took care of one another. You never questioned it. In our family we never expected anything in return. It was the honor that we had. The trust —that's why we all felt so close to each other. There was no house divided here."[27]

The strong feelings evoked by struggling parents in their working-class children led not only to a deep-seated desire to assist the family of origin, but also to a resolve to establish families of their own; life was to be lived within the context of marriage and family. Tied closely to their industrial economy, working-class community, and kin, working class children saw adulthood in the 1920s and 1930s as an extension of their familial and communal culture, not as a departure, although a marriage of one's own did represent something of an independent thrust. Daughters and sons grew to emulate their parents and maintain the system of sharing and assisting. Jule Saniecki respected her mother immensely for raising her despite the tragedy of losing three children to the flu, and wanted to raise a family of her own. Stacia Feddock detailed how her mother struggled to raise five dollars for a first communion dress, and then revealed that her own life would have been "unthinkable" without a child. Another woman exclaimed:

> That was the social thing to do. You didn't say you would get married and not have children. That wasn't the pattern. I'm glad there was a pattern.[28]

Indeed, grandchildren themselves could represent an additional contribution on the part of children to their admired parents. Marie Skovranski and her husband revealed how her father rejoiced at the birth of her son because the infant replaced, in the grandfather's eyes, Marie's brother who had died at age seven.[29]

While marriage was eagerly pursued, working-class family ties and patterns of carrying and assisting each other were not discarded; children continued to care for their parents, especially as they reached old age. After Stacia Feddock's father suffered a serious stroke, she would sit with him daily and listen to stories about his life in Poland. Even when Virginia Vida reached middle age there was never a day that she did not visit her mother. Seven children confessed to their inability to move from the coal region as they matured because they felt obligated to remain near their parents. Many took their parents into their own homes or moved in with them to offer care during old age. Vincent Zaniecki's brother actually built his parents a home. Marriages were postponed past age forty, or even fifty, so that care could be extended for elderly parents. Marie Siegley visited her parents every day until they died. "After they died I don't even remember the times," she said, "it was like a blackout."[30]

But it would be a mistake to conclude that reciprocity did not exist in parent-child relationships. Parents knew all along that they would need both the material and emotional support of their children, and provided both the discipline and the valuable service which would insure child support in subsequent years. Parents were not about to leave the possibility of child assistance to chance or sentiment alone. When the situation called for it, they were not adverse to insisting upon child labor and earnings. When Marie Skovaranski was in the eighth grade, her brother became ill and her mother asked her to leave school and help care for him. After her father died of a heart attack, Virginia Vida's mother "made the kids work." At age sixteen Virginia went out to do house work and baby-sit.[31]

Often a child's work efforts were directed toward the small family business pursuits that were established throughout the industrial regions, including the anthracite coal fields. Arlene Golobek's parents opened a small grocery store in Nanticoke, hoping that the business would ultimately enable her father to leave the mines. Arlene was kept busy seven days a week in the enterprise, cleaning mushrooms and helping in the store. Eleanor Ostrowski left school after the ninth grade and worked in her parents' store.[32]

Children, of course, did not always appreciate strict parental discipline or demands to enter the workplace at an early age and relinquish wages. Arguments occurred frequently. Several young men ran away from home because they had to work for their father. Discipline was often enforced with a strap. One woman resented her father's attempt to regulate her dating patterns and influence her choice of a marriage partner. Sophie Reakes described how children in her family had to kiss their parents' hands and ask for forgiveness before they went to Holy Communion in the Catholic church.[33] Clearly, parents felt that the family could not act in concert to meet economic exigencies over a lifetime without a certain degree of discipline.

But discipline was not all parents had to give. They also proved indispensible in providing housing, a valuable commodity in coal regions, even after children had been married. When they were unable to afford housing of their own, young couples moved in with their parents, often for several years. Stacia Feddock's parents saved feverishly for a home and then acquired a second one next door so they would be able to give each of their two daughters a home of her own. One woman reported that she and her husband moved in with her parents whenever he was out of work. Another man revealed how he lived with his mother-in-law for awhile after marriage and then built a home of his own on a lot he inherited from his mother. Regina Haranzy's father received a home from his parents, which they built on the back of their lot; the parents continued to reside on the front part of the property. When one miner got sick and could no longer repair his home, he asked his son-in-law to take it from him for one dollar and keep it in repair. While houses were often built very cheaply by family members themselves, the widespread sharing of housing resources within family systems underscored the intricate workings of the family economy.[34]

Despite a general pattern of shared responsibilities within these working-class families, some members exerted more authority than others. Certainly males exercised some authority and discipline; unable to derive status from their menial jobs, it might be expected they would seek status at home. But it is necessary to point out that both their status and authority at home were usually secondary to their wives. Women were simply pre-eminent at home and assumed a wide range of responsibilities including management of family finances. In many instances it was coal-region housewives who decided to initiate small family businesses or purchase a home. This does not seem surprising since women generally possessed the most accurate knowledge of family financial resources, which were regularly turned over to them by husbands and children. As Helen Hosey (and nearly every respondent in this study) explained, "My mother always handled the money; my father never even opened his pay envelope."[35]

Women took their responsibility seriously and pursued their role aggressively. When Sophie Reakes spent one dollar of her earnings on a petticoat before turning her wages over, her mother threatened to throw her out of the home. Women frequently risked social ostracism to enter local taverns on payday before husbands had an opportunity to drink away modest earnings. Many miners never entered taverns at all, but those who did posed a threat to family harmony and economy. Indeed, one woman recalled hiding wives in her family's store who feared additional beatings from drunken spouses.[36]

Whatever the role one assumed within the working-class family, all its members lived with certain pressures and tensions. Sharing responsibility may have proved functional but it was not necessarily an easy thing to do. Most working people, of course, were able to meet their familial and economic obligations, but frequently many could not and lapsed into periods of emotional stress. What seems essential to an understanding of the working-class family economy is the types of mental illness which predominated in areas such as the anthracite region. Not everyone, of course, manifested a form of psychosis, but those who did provide clues to the types of emotional stress these families endured and reveal the tensions which all faced whether they could deal with them or not.

Fortunately for historians, Retreat State Hospital for the mentally ill sat in the middle of the anthracite coal fields. The admission records of this institution have survived and offer an opportunity to probe inside the confines of the family economy which was so instrumental in shaping the behavioral and additudinal patterns of mining families. Nearly all the patients admitted to Retreat in the period under study (1920–1940) were first- or second-generation Americans, with over seventy per cent from Poland and Czechoslovakia. Admission registers also revealed the marital status of each patient, his age, and the particular form of psychosis.

In sampling every third admission at Retreat from the late 1920s to 1940, it is apparent that certain psychoses predominated among the mining population. In a period when immigrant parents, largely from eastern Europe, and their progeny were dealing with crucial questions of sharing responsibilities and confronting hard times, the two most frequent forms of psychoses were dementia praecox, an early term for schizophrenia, and manic depression (see table I). Among male admissions between 1929 and 1940, 49.6 per cent were classified as having dementia praecox, a figure which would probably have been higher if contemporary perceptions and diagnostic measures were in force at this time.[37] Among females, the largest group suffered from manic depression—over forty per cent. What is significant about these rates is that they were much higher than national averages for the period. The male dementia praecox rate was more than twice the national rate for 1937 and 1939, which was twenty-two per cent of all admissions. The female incidence of manic depression was about three times the national average of the late 1930s. Even when sex was controlled for, anthracite-region females had a rate of manic depression three times that for females nationally, and the men's rate was three times that of all men.[38]

It is even more instructive to look at trends in the admissions at Retreat as economic conditions deteriorated and families felt increased pressures after

TABLE I
DIAGNOSIS BY SEX:
Retreat State Hospital, 1929–1940 and National Averages*

	Male	Female	Total	National Average 1937	National Average 1939
Dementia	49.6%	41.9%	40.6%	22.2%	22.4%
Manic	17.5%	40.3%	28.1%	13.5%	11.9%
Alcohol	18.2%	3.0%	11.3%	6.0%	5.1%
Other	16.0%	24.3%	20.0%	—	—
N	218	197	416		

*Retreat State Hospital, Register, 1929–1940, Pennsylvania Historical and Museum Commission, Harrisburg. National averages are cited in Leopold Bellock *Manic Depressive Psychosis and Allied Conditions* (New York, 1952), 21–32.

1930. Between a sample taken for 1929 and 1930 and one from 1933 and 1934, during the midst of unemployment and labor strife, the high rates of dementia and manic depression rose even higher. Dementia or schizophrenia, an affliction frequently stemming from disjointed parental-child relationships, climbed from 25.8 per cent to 30 per cent of all admissions for females. Among men the rates increased from 48.4 per cent in the 1929–1930 sample to 54.9 per cent in 1933–1934, a figure that was an astounding five times the national average for all males. Manic depression rose from nine to fourteen per cent for male admissions in the period, but rose from 38.7 to 46 per cent among females, a rate four times the national average for females.[38] Manic depression, while not originating in familial relationships, could very well have been triggered by difficult economic, familial, or work situations among individuals already predisposed by heredity to deal with these problems poorly. Interestingly, admissions for alcohol-related problems dropped for male and female workers during the period, a fact which could have been due to improved diagnostic methods.

Further refinement of the data permits an understanding of intergenerational differences within mining families. Throughout the period, second-generation (born in America) males displayed higher rates than their fathers. Almost sixty per cent of immigrant sons admitted to Retreat in 1933–1934 were suffering from dementia compared to forty per cent of their fathers. Since schizophrenia in males is frequently linked to narcissistic and troubled mothers, we might expect to find troubled women in the first generation. As ex-

TABLE II

DIAGNOSIS BY SEX:

RETREAT STATE HOSPITAL, 1929–1934*

	Female		Male	
	1929–30	1933–34	1929–30	1933–34
N	50	81	51	116
Dementia	25.8%	30.0%	48.4%	54.9%
Manic	38.7%	46.0%	9.0%	14.0%
Alcohol	6.0%	4.0%	15.0%	11.2%

*Source: Retreat State Hospital, Register, 1929–1934, Pennsylvania Historical and Museum Commission.

pected, manic depression (1933–1934) among foreign-born mothers ran sixteen per cent above the rates for their daughters during the same period. It must be admitted that dementia tends to manifest itself at a relatively young age and thus the rates of incidence for native-born males should exceed those of their immigrant fathers. But manic depression is not so age specific and the accelerating rates among foreign-born mothers between 1929 and 1933 suggest women felt increasingly pressured by their circumstances and responsibilities within the family economy as economic conditions worsened. By 1939 the rate of manic depression among foreign-born female admissions was an astounding seventy per cent, while it had decreased among their daughters, who now suffered more from illnesses like dementia, which originated in relationships with their troubled mothers and less in objective social and economic circumstances.

The central point emerging from this analysis of psychoses and working-class family behavior is the suggestion that females who played pivotal roles in family management had a relatively high tendency toward manic depression. Hard times would strike at them directly as administrators of family finance. Both parents would inevitably be troubled if lack of work or income threatened their ability to provide their portion of the shared economy. While an understanding of manic depression has expanded since 1930, it seems increasingly apparent that while its ultimate origins reside in chemical imbalances in the brain,[40] it is initiated by environmental conditions such as financial troubles, sexual difficulties, stress, bad home situations, or even menses in women.[41] More often a woman than a man, the manic patient usually exhibits emotional extremes of excitement and depression—"mood shifts inappropriate for objective circumstances." Loneliness, an inability to forget problems for a

TABLE III

DIAGNOSIS BY GENERATION AND SEX:

Retreat State Hospital, 1929–1940*

	Men			
	Foreign Born		Native-Born	
	dementia	manic	dementia	manic
1929–30	40.0%	6.0%	52.6%	10.5%
1933–34	40.3%	7.0%	59.5%	17.0%
1939–40	11.0%	50.0%	43.4%	26.0%

	Females			
	dementia	manic	dementia	manic
1929–30	7.6%	46.1%	39 %	36.8%
1933–40	34.0%	55.5%	21.2%	39.3%
1939–40	11.7%	70.0%	51.7%	20.6%

*Source: Retreat State Hospital, Register, 1929–1940, Pennsylvania Historical and Museum Commission.

while, and even suicidal tendencies can all be found in the manic patient.[42]

Certainly, objective circumstances existed between 1929 and 1934 in the hard-coal regions to heighten the already elevated levels of manic depression in women who were predisposed toward the disease by heredity. But this fact indicates that pressures were being faced even by stronger women who could more readily withstand them. It also underscores the central responsibility most females attached to their roles within the family economy and how important they felt it was to be able to do their part. As they suffered, however, they ran the risk of emitting contradictory emotions toward their children. Certainly they cared for them and expressed love. This is evident in the recollections cited earlier where children expressed a deep sense of loyalty to their parents. But the family economy also depended upon the imposition of a high degree of discipline; children had to be made aware that they had responsibilities to contribute and share as well as their parents. Most of them learned this lesson well. But the burden remained and, indeed, intensified during hard times as parents provided both love and stern discipline. These emotional signals, however, often could be seen as contradictory in the eyes of the child and therein lie the origins of schizophrenia, especially in the sons of these families who not only had received the conflicting messages but often represented in the eyes of a narcissistic mother the family's best hope to stablize their frequently unpredictable economic circumstances.

Dementia praecox or schizophrenia is a complex disorder, but a consensus seems to exist as to its origin. More commonly found in sons, the classic case stems from a situation in which a mother feels insecure and unfulfilled and attempts to secure her major emotional gratification through her children, especially the son. The son receives conflicting signals of intrusiveness and imperviousness. That is to say, the mother is impervious to the child's needs as an individual (something always possible in the family economy) and simultaneously intrusive in attempting to direct the child's life to her own ends. This situation is exacerbated by passive fathers—a common phenomenon in these mining families—who fail to provide emotional satisfaction to their spouses and traditional images of masculinity to their sons.[43] The mother also is responsible for schizophrenia in the female child when she conveys hopelessness and meaninglessness to her daughters, because she is disappointed in herself or contemptuous of her husband. Sons (second-generation) of mining families showed steadily rising rates of dementia praecox as mothers became more troubled between 1929 and 1934. When hard times eased after the mid-1930s, the rates of dementia in second-generation males declined slightly, although they were well above national levels. Second-generation female rates for dementia did decline between 1929 and 1934 but rose to unprecedented levels by 1939, an indication that as daughters were maturing in this milieu, they were manifesting the pressures of earlier tensions in the family economy. That national studies have consistently demonstrated higher rates of schizophrenia in lower-status occupation groups in industrial regions only underscores the tremendous premium placed on shared responsibility, interaction, and proper fulfillment of roles within the working-class family economy.[44]

If expectations were high for respective family members to perform prescribed roles, and resources such as wages and housing had to be shared, a collective approach to life also extended into the working-class enclave. The gathering of food was a continuous process which required participation from extended kin and neighbors as well as parents and children. Men would prepare smoked sausage and women would make sauerkraut, can fruits and vegetables, and store them in cold cellars. Neighbors would even swap foods for greater variety. Feathers were peeled from ducks for pillows and blankets. And everyone picked coal to enhance winter supplies.[45]

On religious holidays extended kin would gather for dinner and visitations. Christmas celebrations were especially remembered as familial and communal affairs in which traditional customs that stemmed from eastern Europe predominated. Ironically, it was the father acting as titular head of the household who passed the sacred wafer ("oplatek" in Polish) around the Christmas eve table. While economic realities elevated the internal status of the mother, tradition and the soothing influence of nostalgia served to mute realities of work and life and made the world appear as it should during holidays, and not as it was. Some children even recalled that they prayed for familial and communal goals, such as better homes for their parents and kin.[46]

Often, communal sharing became much more demanding. Many women would actually raise the children of female neighbors who died. Mildred Zejack was raised by an aunt after her mother died in childbirth. Frequently, older brothers and sisters raised younger ones when parents passed away. Viola Mazza's parents took in a neighbor's daughter of five when her mother died. Even with ten children of her own, Lillian Niziolek's mother would bring food to neighborhood women who had given birth. Sophie Reakes recalled that even though her mother had eleven children, she took in another who was orphaned. During weddings and funerals, large numbers of neighbors would gather and give food and comfort. Chester Brazina revealed that during funerals, neighbors would bring hams, care for smaller children, and even take time from work to sit in the homes of the bereaved and lend emotional support. Men would pick extra coal for neighbors who were unable to work.[47] And every mining family relied on the cooperative spirit of the local storekeeper who allowed them to buy "on the book." One miner revealed,

> The fellow that should have a monument is the small storekeeper; he is the guy that kept you on the book. Otherwise you'd never won a strike here. . . . when a strike came—why he just kept on taking on his book and you didn't pay him until you started back to work. If he had demanded his money when it was due, you couldn't have survived.[48]

Certainly the working-class family and community should not be romanticized. Factionalism in the ranks of miners represented by the UMW and the UAM cannot be ignored, and tensions between parents and children were frequent enough; limits to cooperation existed. But this study indicates overwhelmingly that a persistent pattern of sharing and cooperation was rigorously enforced. It was pursued with such rigor, in fact, that the responsibilities of proper role fulfillment could prove difficult to meet for many family members and lead to exceptionally high incidences of manic depression and schizophrenia.

When economic difficulties mounted in the anthracite region after 1930 and miners sought to solve their problems, it should not be surprising that they turned to a solution such as job equalization. They were willing to share available work opportunities because cooperation and sharing were the means by which they had always met the exigencies of industrial life. Above all else, their behavior and perceptions were rooted in the central institution which emerged at the intersection of traditional and industrial cultures—the family economy. They vigorously attacked the United Mine Workers with all their collective resources—men, women, and children—because it stood in the way of achieving their fundamental goal of sharing. While the United Anthracite Miners may have lost its ultimate battle to an entrenched union and the powerful coal companies, the quest for cooperation was genuine, was participated in widely, and was shaped not by repressive companies, big government, or even an industrial workplace, but from below in the intimate workings of the family culture. And in the 1936 anthracite contract, the UMW finally included a statement in support of equalization even though the concept was never implemented.

But the ultimate import of the struggle of anthracite-region people in the 1930s was the powerful statement they made about their priorities. Labor historians who reduce working-class behavior to a product of the workplace and whose major concern is with who exercises greater control on the production line, have unnecessarily reduced the framework in which worker behavior should be analyzed. Industrial workers were not simply, as many Marxist historians would suggest, tied to the material realities of the workplace. They were not, moreover, inherently radical, nor were their views moderated only by governmental intrusion. There was some of this to be sure. But the attitudes of most workers were rooted, ultimately, in the loving concern which united both family and neighborhood and also in the tensions which pervaded these relationships. If industrial managers were never able to impart the degree of discipline on the line they would have liked, it may have been because a stronger dose of discipline was administered at home; the scheme of mutual obligations between parents and children was indelibly impressed on workers. This family economy suffered traumas to be sure; mothers especially exhibited the strain and their sons suffered the consequences. But this only seemed to underscore how seriously family members took their responsibilities. They did not forget that responsibility to each other when the time came to speak out.

NOTES

[1] Staughton Lynd, "The Possibility of Radicalism in the Early 1930's: The Case of Steel," *Radical America,* 6 (November-December, 1972); Alice and Staughton Lynd, eds., *Rank and File: Personal Histories by Working-Class Organizers* (Boston, 1973), 1–7; Harry Braverman, *Labor and Monopoly Capital: The Degradation of Work in the Twentieth Century* (New York, 1974); David Montgomery, *Workers' Control in America* (Cambridge, England, 1979), 164–65; James R. Greene, *The World of the Worker* (New York, 1980), 172; Nelson Lichtenstein, "Auto Worker Militancy and the Structure of Factory Life, 1937–1955," *Journal of American History,* 65 (September 1980), 352–53; Peter Friedlander, *The Emergence of a UAW Local, A Study in Class and Culture, 1936–1939* (Pittsburgh, 1975). Ronald Schatz argues that leaders in the union

drives of the 1930s were skilled workers who were "loosely supervised" at work; see Schatz, "Union Pioneers: The Founders of Local Unions at General Electric and Westinghouse, 1933–1937," *Journal of American History,* 66 (December, 1979), 595. For a view that locates labor attitudes in preindustrial culture, see the brilliant discussion by Herbert Gutman, "Work, Culture, and Society in Industrializing America, 1815–1918," *American Historical Review,* 78 (June 1973).

[2] Steve Nelson, James R. Barrett, and Rob Ruk, *Steve Nelson, an American Radical* (Pittsburgh, 1981), 160–173.

[3] *United Mine Workers Journal,* XLV (15 June 1934), 14; (1 May 1934), 13; *Wilkes-Barre* (Pa.) *Record,* 21 November 1933, 14; *New York Times,* 21 August 1933, 27; 18 November 1933, 7.

[4] "Report on Operations of the Lehigh Navigation Coal Company, Aug. 8, 1938," Lehigh Coal and Navigation Company Records, MG31, Pennsylvania Historical and Museum Commission (PHMC), Harrisburg. This report admitted that labor has learned to act as a unit in resisting efforts of the management to decrease costs. *UMW Journal,* XLV (15 January 1934), 8; 1 May 1934, 4; *Wilkes-Barre Record,* 18 November 1933, 14; 22 November 1933, 1.

[5] Joe Sudol interview by John Bodnar, 10 March 1981 tape recording; Stanley Salva interview by Bodnar, 19 March 1981; Ben Grevera and Anthony Piscotty interview by Bodnar, 24 March 1981 (Pennsylvania Historical and Museum Commission).

[6] Douglas K. Monroe, "A Decade of Turmoil: John L. Lewis and the Anthracite Miners, 1926–1936," (unpublished Ph.D. dissertation, Georgetown University, 1976), 215–30.

[7] *Wilkes-Barre Record,* 18 November 1933, 1, 14; *UMW Journal,* XLVII (August 8–15, 1936), 5; XLIV (1 December 1933), 3–4, 8. See also Pierce Williams to Harry L. Hopkins, 10 January 1935, Harry Hopkins Papers, Franklin D. Roosevelt Library.

[8] An indication of the UMW's lack of interest in the rank and file goal of equalization came at the 1935 UMW convention for the anthracite region. The only resolutions passed at the convention were against the United Anthracite Miners, a rival group, and "pusher bosses." No mention was made of equalization. See *UMW Journal,* XLVI (15 December 1935), 3; Melvyn Dubofsky and Warren Vantine, *John L. Lewis: A Biography* (New York, 1978).

[9] The culmination of this violence came in 1935–1936 when Thomas Maloney, head of the United Anthracite Miners, was arrested along with other leaders of the union for contempt in their failure to honor a strike injunction. Thousands of miners encircled the courthouse during the ensuring trial, which was broadcast on local radio. In April 1936, Maloney and his five year old son were killed when a package which arrived in their mail exploded. See *Wilkes-Barre Record,* 11 April 1936, 1; 8 February 1935, 17; 11 February 1935, 1. 5 March 1935, 1; "Investigation of Reports Concerning Conduct of State Police in Strike Zone, May 6, 1935," Records of the Pennsylvania State Police, R.G. 30, Box 3, Pennsylvania Historical and Museum Commission.

[10] Chester Brazina interview by Bodnar, 15 April 1981, tape recording (Pennsylvania Historical and Museum Commission).

[11] Joseph Krievak interview by Bodnar, 19 March 1981; John Sarnoski interview by Bodnar, 10 March 1981; Vincent Zaniecki interview by Bodnar, 5 February 1981; Salva interview 19 March 1981; all interviews tape recorded (Pennsylvania Historical and Museum Commission). *Wilkes-Barre Record,* 22 November 1933, 1. On the National Miners Union, see Federal Mediation and Conciliation Service Dispute File, R.G. 280, National Archives, Thomas Davis to H. L. Kerwin, 8 April 1931.

[12] Leonard Wydotis interview by Bodnar, 3 March 1981, tape recording (Pennsyl-

vania Historical and Museum Commission). By 1933, the union local at Wanamie to which Wydotis belonged was so strong for the UAM that John L. Lewis revoked their charter; see *Wilkes-Barre Record,* 21 November 1933, 20.

[13]*Wilkes-Barre Record,* 18 November 1933, 1.

[14]George and Stacia Tresniowski interview by Bodnar, 3 March 1981, tape recording (Pennsylvania Historical and Museum Commission). See also Grevera and Piscotty interview, 24 March 1981.

[15]William Everett interview by Bodnar, 29 May 1981, tape recording (Pennsylvania Historical and Museum Commission).

[16]*Ibid.*

[17]For accounts of women participating in anthracite labor protest in earlier times, see Victor Greene, *The Slavic Community on Strike* (Notre Dame, Indiana, 1968) and Michael Novak, *The Guns of Lattimer* (New York, 1978). For an account of unemployment councils in the region, see Nelson, Barrett, and Ruck, *Steve Nelson: American Radical.*

[18]*Wilkes-Barre Record,* 22 February 1935, 13; 1 March 1935, 26; "Industrial Disturbance, Anthracite Coal District," Strike Reports, Pennsylvania State Police Records, RG30, Pennsylvania Historical and Museum Commission.

[19]*Wilkes-Barre Record,* 18 March 1935, 13; 26 March 1935, 13; George and Stacia Tresniowski interview, 3 March 1981; "Arrest for Dynamiting in Luzerne County," Strike Reports, Pennsylvania State Police Records, Box 3.

[20]See Peter Gerko (Plymouth, Pa.) to Franklin D. Roosevelt, 24 March 1935; Lana Gorgus to Franklin D. Roosevelt, 25 March 1935; "Mine Workers of Wanamie" to Franklin D. Roosevelt, 20 February 1935, Federal Mediation and Conciliation Service Dispute File, RG 280, National Archives. Glen Alden president William Inglis sent a telegram to the U.S. Department of Labor stating that he planned to go ahead with the eviction program because UAM sympathizers had interfered with production and the implementation of the contract between the company and UMW. Evidence also existed that companies like Glen Alden were violating the contract themselves by hiring independent gangs of labor in 1932 and paying them below union scale to mine coal. Most men felt this was taking jobs away from their communities. See William Inglis to Hugh Kerwin, telegram, 23 February 1935; Stephen Matyleurs to Francis Perkins, 2 June 1935, Federal Mediation and Conciliation Service Dispute File.

[21]Antoinette Watchilla interview by Angela Staskavage, 12 July 1977; Helen Knepp interview by Bodnar, 16 September 1978; Viola Mazza interview by Staskavage, 18 September 1977; Sophia Reakes interview by Staskavage, 13 July 1977; Adelle Winarski interview by Staskavage, 1 August 1977; Irene W. interview by Bodnar, 26 March 1981.

[22]Irene W. interview, 26 November 1977; Stacia Tresniowski interview, 3 March 1981.

[23]Steve Stashak interview by Bodnar, 19 March 1981; Peter Byczkowski interview by Bodnar, 18 May 1981; Steve Gotcha interview by Bodnar, 18 May 1981; Joseph Sudol interview by Bodnar, 4 March 1981; Joseph Molski interview by Bodnar, 4 March 1981. All interviews tape recorded (Pennsylvania Historical and Museum Commission).

[24]Lillian N. interview by Staskavage, 13 July 1977; Joseph Krivak interview by Bodnar, 19 March 1981; Stanley Salva interview by Bodnar, 19 March 1981; Vincent Znaniecki interview by Bodnar, 5 February 1981; Viola Mazza interview, Berry Hoskins interview by Staskavage, 18 January 1978. All interviews are tape recorded (Pennsylvania Historical and Museum Commission).

[25]Stanley Salva interview, 19 March 1981; Helen Klish interview, 19 September 1977.

[26]May Gill interview by Staskavage, 25 March 1977; Antoinette Warchilla interview, 12 July 1977; Virginia Vida interview.

[27]Helen Hosey interview by James Rodechko, 21 July 1977, tape recording (Pennsylvania Historical and Museum Commission).

[28]Jule Znaniecki interview by Staskavage, 7 July 1977; Stacia Feddock interview by Shirley Donio, 27 August 1977; Marie Siegley interview by Staskavage, 11 January 1978. All interviews tape recorded (Pennsylvania Historical and Museum Commission).

[29]Marie Skovranski interview by Staskavage, 16 January 1978, tape recording (Pennsylvania Historical and Museum Commission).

[30]Stacia Feddock interview, 27 August 1977; Irene W. interview, 26 November 1977; Sophie Reakes interview, 13 July 1977; Pearl C. interview by Bodnar, 17 March 1981; Marie Siegley interview, 11 January 1978, tape recording (Pennsylvania Historical and Museum Commission).

[31]Marie Skovranski interview, 16 January 1978; Virginia Vida interview, 6 December 1977; Joseph Sudol interview, 3 March 1981; Stanley Salva interview, 19 March 1981; Joseph Molski and Leonard Wydotis interview, 10 March 1981.

[32]Arlene Gobolek interview, 11 December 1977; Eleanor Ostrowski interview, 11 November 1977; Sophie Reakes interview, 13 July 1977.

[33]*Ibid.,* Sophie Wojcik interview by Staskavage, 14 September 1977; Stella K. interview by Bodnar, 18 May 1981.

[34]See Stacia Feddock interview, 27 August 1977; Marie Skovranski interview, 16 January 1978; Vincent Znaniecki interview, 5 February 1981; Virginia Vida interview, 6 December 1977.

[35]*Ibid.,* Viola Mazza interview, 17 September 1977.

[36]Arlene Gobolek interview, 11 December 1977; Sophie Reakes interview, 17 July 1977; Helen Klish interview, 17 September 1977; Steve Gotcha interview, 18 May 1981; John Sarnoski interview, 10 March 1981.

[37]Studies have shown that the incidence of dementia praecox diagnosed in the 1930s was particularly low—perhaps by one half—because of inferior knowledge which existed about the disorder. See Judith B. Kuriansky, "Trends in the Frequency of Schizophrenia by Different Diagnostic Criteria," *American Journal of Psychiatry,* 134 (June 1977), 631–34. I am grateful to Professor Thomas Oltmans of Indiana University for this reference.

[38]For national averages see Leopold Bellock, *Manic-Depressive Psychosis and Allied Conditions* (New York, 1952), 21–32.

[39]Numerous factors can induce schizophrenia and a predisposition to it, according to contemporary research, may be transmitted genetically. See Bellock, *Manic-Depressive Psychosis and Allied Conditions,* 30–31.

[40]P. L. McGeer and E. G. McGeer, "Chemistry in Mood and Emotion," *Annual Review of Psychology,* 31 (1980), 273–307; John C. Campbell, *Manic Depressive Disease: Clinical and Psychiatric Significance* (Philadelphia, 1953) 166–171; Horatio M. Pollock, Benjamin Malzberg, and Raymond G. Fuller, *Hereditary and Environmental Factors in the Causation of Manic-Depressive Psychoses and Dementia Praecox* (Utica, N.Y., 1939), 460; Leopold Belloc, *Manic-Depressive Psychosis and Allied Conditions* 21–32. I am indebted to Professor George Rebec, a psychologist at Indiana University for critical comments on my treatment of schizophrenia and manic depression.

[41]Campbell, *Manic-Depressive Disease,* p. 167; Pollock, Malzberg, and Fuller,

Hereditary and Environmental Factors in the Causation of Manic-Depressive Psychoses and Dementia Praecox, 460. Studies have generally found a higher incidence of manic depression among foreign-born rather than native-born in the United States; see Horatio M. Pollock, "Prevalence of Manic-Depressive Psychoses in Relation to Sex, Age, Environment, Nativity and Race," in *Manic-Depressive Psychosis* (Baltimore, 1931), 655–667.

⁴²John R. Cavanagh and Jones B. McGoldrick, *Fundamental Psychiatry* (Milwaukee, 1962), 268–70; Hattie N. Smith, "A Study of the Neurotic Tendencies Shown in Dementia Praecox and Manic-Depressive Insanity," *Journal of Social Psychology,* IV (February, 1933), 116–28.

⁴³While some recent evidence on the influence of genetic transmission of schizophrenia has tended to weaken the case for family relationships as a factor, even recent studies continue to suggest that environmental factors such as mother-child relationships are highly likely to trigger the illness. Even some who have argued that there is a genetic component to the disorder conceed that nongenetic environmental factors are required for the development of the illness. See Frank Summers and Froma Walsh, "The Nature of the Symbiotic Band Between Mother and Schizophrenic," *American Journal of Orthopsychiatry,* 47 (July 1977), 484–86; S. Kety, D. Rosenthal and P. Wender, "Mental Illness in the Biological and Adoptive Families of Adopted Schizophrenics," *American Journal of Psychiatry,* 128 (1971), 302–306; Theodore Linz, *The Origins and Treatment of Schizophrenic Disorders* (New York, 1973), 48–49, 67–69; Silvano Arieti, *Interpretation of Schizophrenia* (New York, 1974), p. 53; Pollock, Malzborg, and Fuller, *Hereditary and Environmental Factors,* 460.

⁴⁴Arieti, *Interpretation of Schizophrenia,* 496, 499.

⁴⁵Joseph Molski interview, 31 March 1981; Leonard Wydotas interview, 3 March 1981; Helen Klish interview, 17 September 1977; Arlene Golobek, 11 December 1977.

⁴⁶Sophie Wojcik interview, 14 September 1977; Helen Klish interview, 17 September 1977; Virginia Vida interview, 6 December 1977; Marie Siegley interview, 11 January 1978.

⁴⁷Mildren Weiss interview, 23 August 1977; Lillian Nizolek interview, 14 September 1977; Joseph Krivak interview, 19 March 1981.

⁴⁸Chester Brazina interview, 15 March 1981. Although it was illegal, even tavern owners extended credit during strikes in hard times.

Chapter Eight

THE COAL AND IRON POLICE IN ANTHRACITE COUNTRY

Stephen R. Couch

The social history of police in this country is a fascinating story which is only beginning to be developed. The dominance of public, governmental police has come about only in the last hundred years, and was accomplished through struggles of differing groups with many competing interests. Prior to the ascendance of public police, various forms of private and quasi-public police existed and were charged with carrying out the bulk of police functions.[1]

One of the most interesting forms of quasi-public police had its birth in Pennsylvania's anthracite region and became the dominant type of constabulary in that region during the last quarter of the nineteenth century. Pennsylvania's coal and iron police existed between 1866 and 1935. Created by an act of the legislature, they were controlled and paid by the coal and iron companies, yet were commissioned by the governor and given full powers of public police. While similar arrangements existed in short-term situations in other states, there is no evidence that such an arrangement was ever as long-lived or as strongly institutionalized as in Pennsylvania.

This paper examines the development and role of the coal and iron police in the anthracite region. It is argued that these private forces developed due to conditions distinctive to areas such as the anthracite coal fields in the nineteenth century; that they were successful in helping to establish and maintain the social and economic dominance of the coal companies during that period; but that in the long run, they were instrumental in provoking successful resistance against coal-company domination, and helped encourage the building of a strong labor organization in the anthracite fields.

Conditions Which Spawned the Coal and Iron Police

The discovery of the existence of coal in Pennsylvania, and of its many uses, dated from the mid- to late eighteenth and early nineteenth centuries. Differences of opinion abound on crediting the original discoveries of various deposits and different uses; but it is certain that by the mid-nineteenth century, coal was essential to the expanding industrial capacity of the U.S..[2] Production rose from one-and-a-half million tons in 1833 to over twenty million tons in 1860.[3]

In the anthracite fields of eastern Pennsylvania, the period from the late 1820s to the early 1870s was one of great individual land speculation. While unstable market and production conditions caused some to urge early consolidation of the industry, the results of the laissez-faire system were impressive overall, with production increasing steadily through mid-century.[4]

However, as the industry developed, a serious problem emerged: chronic overproduction and over-investment.[5] When demand, and hence prices, fell sharply after the Civil War, the situation became acute. Various railroad companies began buying coal lands in order to control production of the product they transported to market. The greatest coal empire of the day became the Philadelphia and Reading when, in 1871 and 1872, its president, Franklin B. Gowen, bought sixty thousand acres of coal land, followed by forty thousand more acres over the next two years. Consolidation was swift: in 1875, fewer than thirty-six of the region's 175 collieries were independently owned.[6] By the early twentieth century, the takeover was even more complete. By 1900, three railroads—the Reading, the Lehigh Valley, and the Delaware, Lackawanna and Western—owned or controlled sixty per cent of the anthracite coal deposits.[7] In 1902, over ninety per cent of the anthracite deposits were owned or controlled by a railroad, with ninety-one per cent of the land being owned outright.[8] So, during the last quarter of the nineteenth century, Pennsylvania coal regions became internal colonies to core areas along the east coast. Coal and profits flowed out of the region, helping to create the development of core cities, institutions, and the urban upper class.[9]

Coal barons and their agents continually attempted to dominate the social and political order. The most overt method of social control was the company town. These unincorporated villages began innocently enough out of geographic and economic necessity.[10] Mines were geographically fixed and were often isolated from existing towns; miners needed housing but couldn't afford to build it; and private builders were not willing to risk the high probability that a new mine would fail after a short time, creating a ghost town and the developer's financial ruin. So the companies built towns.

Once created, the towns offered tremendous opportunities for the companies to control their work forces. Movement in and out of town was controlled; unruly miners were evicted; rents were deducted directly from miners' pay, as were items bought from company stores at which miners and their families were forced to shop, often paying exorbitant prices. Even priests' salaries were deducted from miners' earnings. Investigations revealed conditions in company towns were poor, relative to free towns.[11] Company towns have often been compared to medieval fiefs.[12] Actually, a colonial analogy is more apt, for in company towns, wage laborers were further subsidizing their employers/absentee landlords through the operation of this closed social system.

Due primarily to geography, company towns were much more prevalent in the bituminous coal regions than in anthracite country. But even in free towns, coal operators were powerful forces.[13] Coal companies were still the largest employers; often, they provided water for the towns; attorneys were kept busy with coal leases and contracts; company doctors and clergy would be influential in providing community services; favored merchants would benefit from company credit arrangements; free railroad passes would be used to court influential local citizens; and charity would be used to the companies' best advantage.

Politically, the companies made their influence felt at all levels of government, using lobbying and patronage to the fullest. Aurand reports that mine officials succeeded in reducing the power of many local offices to nearly nothing.[14] Yearley points out that the Commonwealth of Pennsylvania had the official power to control the Reading's voracious expansion, but did not, because, in fact, the Reading constituted the State's "interim government."[15] The most blatant illustration of the expropriation of State sovereignty by the companies is the famous Molly Maguire case in the 1870s. Private detectives hired by the Reading investigated the case; company police arrested the alleged offenders; and Franklin B. Gowen, Esq., prosecuted the case himself. As Aurand quips, "The State provided only the courtroom and hangman."[16]

The Mollies are only one indication that social conditions in the coal regions were very volatile. Throughout this period, a "cultural division of labor" was present, whereby certain types of jobs were assigned to certain ethnic groups.[17] Ethnic divisions and antagonisms appeared early, were altered by the entrance of new groups, but were constantly a point of conflict within Pennsylvania's coal regions.[18] Ethnic divisions created a significant amount of social disorder throughout the region; unified class or community interests failed to develop.[19] Operators played one ethnic group off against another, making it difficult for miners to take successful unified action against the absentee owners. One indicator of the resulting problems of disorder was that in 1864 the legislature passed a law banning the carrying of concealed deadly weapons in Schuylkill County.[20]

Some of this disorder had ethnic causes; some was a by-product of a frontier-like social environment; but the roots of the problem were found in labor issues. Strained relations between workers and coal companies existed even before the Civil War. After the war, labor troubles became acute. The companies needed to find a solution to the problem of labor control in areas devoid of reliable and numerous local police.

Advent of the Coal and Iron Police

By the mid-1860s, finding a solution to social disorder and unrest in the anthracite fields had become a matter of concern to the State legislature. A number of bills were introduced in 1866 and 1867 which would have provided for a uniform police system in Schuylkill and Luzerne counties. While differing in detail, these bills attempted to develop a system of public police accountable to the Commonwealth of Pennsylvania (either directly to the governor and/or to commissioners appointed by the governor), thus by-passing local directly elected officials who were seen by coal operators and their allies to favor the miners.

Debate on these bills was many-sided and sometimes passionate. Even the extent, and certainly the cause, of lawlessness was held in question. One representative argued that the lawlessness was confined to an area of less than six square miles, and that the miners could not be blamed for the outlawry.

> This state of things, Mr. Speaker, is brought about by the landholders themselves. . . . They have aggravated those men to desparation, and taken from them their lease rights of land on which some of them had built houses, worth from one thousand to three thousand dollars. . . . [As for lawlessness,] we can provide a cure for the disease wherever we find it. We will put it into the hands of our courts, and hold them responsible until the evil be remedied."[21]

Other legislators thought much more severe measures were needed. As one put it:

> We should convince these ignorant, classbound, infatuated, reckless men that the executive power of the whole Commonwealth has been involved to enforce the execution of law and to punish crime.[22]

Or, in an eloquent fashion which refers to the attack on U.S. Senator Charles Sumner by "Bully" Brooks on the floor of the U.S. Senate in 1856:

> This violence in Schuylkill County must have its death warrant. It must be settled. . . . The spirit that Brooks exhibited towards Sumner is the . . . spirit, the murderous spirit upon the wild horse of crime in Schuylkill County. Brooks was the personification of a two-thirds-grown rebel whelp, when Democracy was three-fourths respectable, and his spirit is now in the coal regions of Pennsylvania, riding upon the unbridled white horse of sin. That horse, that white horse of death, must be captured. The

strong curb of the law must be put in his mouth, and a halter with a noose put about his neck."[23]

So the debate raged on, with no concensus emerging over exactly how the "white horse of death" was to be controlled. Some thought the state would gain too much power with a state-controlled police force. Some thought it was unfair to single out certain counties for employment of such a force. Some felt that county residents should be taxed to pay for special police, while others argued that the coal companies should pay, since the police would be of most benefit to them. In the end, all of these bills failed.

In the meantime, however, the companies and their supporters had found another means of creating a constabulary supportive of their wishes, a force which would have the full police powers of the State while being controlled directly by the coal companies. This came in the form of a supplement to an 1865 law which allowed railroad companies to employ police commissioned by the governor to "exercise all the powers of policemen of the city of Philadelphia."[24] In other words, railroad police could arrest suspected offenders anywhere within the several counties in which they were commissioned. This made sense in that it allowed these policemen to pursue felons on land which was not contained in narrow railroad right-of-ways.

This same logic does not apply to the coal and iron police, who were given identical powers by supplementary legislation passed in 1866.[25] Through this law, coal companies gained a tremendous weapon by which to attempt to control the population of the coal regions. Companies needed only to submit the names of individuals they desired to employ to the governor, who would grant the commissions. No time limit was set on the life of the commission. The governor had the power to revoke commissions, but there is no record of a governor doing so until the 1920s, nor is there any evidence that the State attempted to learn any background information about people it commissioned.[26] In 1871, the fee of one dollar was charged by the State to the companies for each commission granted.

Shalloo summarized the situation:

> In substance, the practice . . . was simply a contract between mining companies and the state whereby police power was "sold" to the industrial establishment. When the commission was issued and the fee paid. . . . the state took no further interest. There was no investigation, no regulation, no supervision no responsibility undertaken by the state, which had literally created 'islands' of police power which were free to float as the employers saw fit.[27]

The coal and iron police represent a solution for business interests to similar problems which plagued society's "respectable elements" in nineteenth-century cities. There, social and labor unrest led to the creation of municipal police forces and then to attempts at police reforms, which would secure control of the police from Democratic party machines.[28] In the coal regions, elected officials (including law enforcement personnel) tended as often as not to side with miners in disputes with coal companies. The companies needed to find a police system which would be responsive to their needs, protect their property, and to create a docile, manageable work force out of the diverse, uprooted immigrant groups arriving in the region.

The problem was compounded by the fact that the coal region had important characteristics which were very different from those found in major cities. Foremost among these was the region's geographic isolation. Instead of growing from a center-city nucleus outward, the region was characterized by a series of towns and mine patches springing up around mines set in the rugged terrain of Pennsylvania's mountains. Transportation systems were developed to get coal out of the region, not to link areas within the region. Added to the physical and ethnic separations (which usually coincided) were political divisions of the area into small, separate units. The only strong controlling force uniting parts of these regions was their dependence on coal operators—many not being from the local area—for their economic existence.

In this context, the development of adequate municipal police departments was out of the question. As with most rural areas, law enforcement forces were organized around a system of county sheriffs and local constables.[29] This system was proving to be not very effective in rural areas. Add to this the urban-like social disorganization of the regions, and public forces became completely inadequate.[30] In addition, county governments were reluctant to pay for additional public police whose duties would consist largely of protecting the property of coal operators, including policing private company towns.[31]

Finally, there was the critical issue of economic and labor control. By the Civil War, coal had become a critical commodity for the increasing industrial development of core areas. It took on added significance during the Union's war effort. Keeping the mines producing and safe from Confederate sabotage (and from labor disorders of northern miners) was a serious preoccupation of operators and the federal government. Troops were sent into the anthracite region to guard the mines.[32] On at least one occasion (12 July 1864), the government took over the mines, but inexperience in mine operations caused the mines to be returned to private management in short order.[33]

Violence and civil and labor disorders were not uncommon even before the War. A boatmen's strike on the Schuylkill Canal occurred in 1835.[34] The first miners' strike in Schuylkill County occurred in 1842 and was dispersed with the appearance of militia.[35] The first miners' union in the county appeared in 1848.[36] Rumors of violence by "Molly Maguires" began in 1857. The area miners protested vigorously when the military draft was instituted in 1863.[37]

By 1866, it was apparent that labor conditions were not stabilizing and that some means of maintaining effective order was required.

A private police army seemed the ideal short-term solution. Cloaked with the power and authority of the State, this force would be controlled directly by, and in the interests of, the operators. During times of civil or labor disorders, the force could be expanded and armed very quickly, then reduced in size when peace was restored—all with the blessing, but not the interference, of the State. For the State and local governments, the solution provided an inexpensive and bureaucratically simple remedy for the law enforcement problems existing in the coal regions. Conceptually, the creation of the coal and iron police is perhaps the most blatant and direct case of police acting as an army of occupation in the interests of colonial powers, which are in this case, not governments, but corporations.

Role of Coal and Iron Police in "Peaceful" Times

The 1860s and 1870s saw the arrival of coal and iron police in anthracite country, increasing in numbers as labor unrest became more widespread.[38] Prior to the passage of Pennsylvania's Mansfield Bill in 1929, no systematic public records of numbers or activities of coal and iron police were kept. What is known must be pieced together from contemporary reports, commission investigations, and a few examples of issued coal and iron commissions.[39]

A few surviving 1869 commissions from Schuylkill and Luzerne counties provide us with clues as to who the early coal and iron policemen were and how they operated.[40] The names of these men show that many were of Welsh, English, and German origin. As would be expected, possible Irish names are conspicuously few. The commissions list a number of companies for whom the police could work, implying that there was a pool of police available to companies when in need, and that some coordinating mechanism between the companies likely existed to deploy and perhaps pay these men. When the railroads consolidated their dominance of the region by buying out most independent coal operators in the early 1870s, this practice of sharing coal and iron police apparently ceased.

During times of labor peace, the number of coal and iron police seem to have been relatively small. However, their duties were many and varied. As with the public police of the time, their orientation was more toward what Wilson terms "order maintenance" than "law enforcement."[41] They patrolled company towns, coal lands, and collieries, attempting to keep order and protect company property.

Another similarity with public police of the day (as opposed to today's police forces) was their social-service concerns, such as inspecting public health and sanitation conditions, and keeping "the company towns free from immorality."[42] But their peculiar structural circumstance of being a kind of private occupation army placed additional burdens upon them. Often they were required to keep "undesirables" out of company towns, and to keep miners in them.[43] They functioned as eviction officers and debt collectors. Indeed, even their mere physical presence as loyal company employees was useful. In the 1870s and 1880s, when major companies were battling for control of coal lands and with possession being nine-tenths of the law, a company would erect a town on disputed land and inhabit the homes with coal and iron police until the matter of ownership was settled in the courts.[44]

The difficulty of carrying out such a myriad of duties, while not being seen as legitimate by most of the local constituency, was exacerbated by the ethnic diversity of the regions patrolled. The ideal coal and iron policeman would have to be familiar with the customs and languages of numerous peoples, show no partiality toward any one group, and possess much patience, sensitivity, and tact. Too few did, and too many simply relied on the gun and blackjack as the ultimate authority.[45]

An additional area of constant tension was the relationship of coal and iron police to local public police officials. Most local government officials may have welcomed the added protection given to their bailiwicks by coal and iron police, but for local constables and sheriffs, these private forces were a mixed blessing.

To be sure, the coal and iron police were used gratefully by local law officers during strikes and other emergencies, often forming the core of posses organized by county sheriffs.[46] But there were seemingly constant problems of jurisdiction. For example, it was not uncommon for coal and iron police to bar public police from entering company towns to serve warrants.[47] And public police and magistrates, who often viewed coal and iron police as competition and something less than "real" police, were not very cooperative. Should coal and iron police arrest a person and take him to the local public constable or to a local magistrate, the prisoner was often summarily released. This led to various abuses, such as doling out private "justice" in the confines of company police barracks, or taking the prisoner to a "friendly" magistrate, often many miles from where the violation occurred. When the latter course of action was followed, it was normal practice to charge the prisoner mileage for the trip.[48]

In any case, ultimate power was clearly on the side of the coal and iron forces, who in effect reduced many local constabularies' influence to next to nothing. Aurand reports that in 1891, the city of Pottsville, with a population of fourteen thousand, employed nine police, and nearby Shenandoah, which claimed nearly sixteen thousand inhabitants, was protected by two public policemen.[49]

The role of the coal and iron police during "peaceful" times, then, was many-faceted. These private police combined the duties of watchman, constable, social service inspector, and keeper of public morality with those of maintaining tight corporate control over company towns, collieries, and their environs. There is no question that they were the dominant law enforcement power in the anthracite region during the last quarter of the nineteenth century. Much of the hatred directed against them can be understood by realizing the enormous difficulty any group would have encountered trying to carry out such varied assignments. Such problems were heightened by these police being direct representatives of the companies and by the inability of many of them to overcome anti-immigrant and ethnic biases of the time. But, as mentioned earlier, it was their role in labor disputes which generated the most controversy, and it is to this subject that we shall now turn.

Labor Troubles

The famous Mollie Maguire incidents marked the coal and iron police's first appearance in force and represent their consolidation of power in the anthracite region.[50] A reputedly secret organization of violent Irish miners and supporters, the Mollies were blamed for a great number of murders and other acts of violence during the 1860s and 1870s. In what Rochester reports to be the first recorded use of labor spies, Reading President Franklin B. Gowen employed the Pinkerton Detective Agency to infiltrate and destroy the Mollies.[51] One Pinkerton agent was made captain of Reading's coal and iron police, while another gained entry to the inner circle of the Mollies and later provided testimony which led to their downfall. Twenty reputed Mollies were executed in 1877–1878.[52]

Coal and iron police also were employed against miners' unions, being used to spy on and intimidate miners to protect strikebreakers and to stop marches and quell (or sometimes cause) riots during strikes. Reading President Franklin B. Gowen once said, "I'll turn Schuylkill County into a howling wilderness before I give in to the miners."[53] Coal and iron police provided the frontline troops in coal operators' efforts to thwart miners' strikes and unions. During the "long strike" of 1875, "special police" were imported from Philadelphia to help defeat the strike and destroy the union.[54] Their efforts were successful.

The pattern of importing large numbers of men from outside the region, commissioning them as coal and iron police, and employing them for the duration of labor hostilities, was repeated throughout the nineteenth century. Attempts to recruit local residents as coal and iron police or deputy sheriffs met with considerable resistance from the populace.[55] When such attempts were successful, most of the population held the recruits in contempt.[56] Therefore, companies needed to recruit men from major cities to supplement the local forces. These "outsiders" were a major irritation to the local populace. An article in the *Pottsville Republican* reported the citizens of Shenandoah as feeling that the mere presence of coal and iron police in the city was likely to incite violence, and while the regular coal and iron policemen were generally tolerated, the extra men brought in to augment the force were regarded as "particularly obnoxious."[57]

Questions on the character of imported coal and iron police abounded. The Anthracite Mine Strike Commission heard testimony from nine coal and iron policemen from Philadelphia:

> They told that they went up into the coal region at the solicitation of representatives of detective agencies to assist in the laudable enterprise of helping to preserve the peace and earn $2.50 a day and free board and lodging. Only one of the nine was asked for a recommendation. . . . As a rule they were sworn in as coal and iron police, furnished with a rifle and put on duty guarding colliery property. Some of them were sworn in as deputies. One of them told that he was in a squad of twenty-eight sent up from Philadelphia; that there were men from Chicago, New York and even Camden in the party, and that these men were made deputies by the sheriffs of Luzerne and Schuylkill counties. . . . One of the witnesses said his occupation was "canvassing from house to house." He did not volunteer the information as to whether it was first or second story canvassing and Mr. McCarthy ventured to ask him if he had ever been locked up. . . . "Oh, maybe I'se been locked up a few times . . . just for havin' a little fun."[58]

The numbers and deployment of coal and iron police varied from strike to strike, and at times was very substantial. For example, in 1888, eight stations of thirty to fifty coal and iron police each were established around Schuylkill County.[59] The next month, with disorder expected in Shenandoah, a sheriff's posse of 285 men and fifty special police sworn in by the chief burgess, were supplemented by two hundred on-duty coal and iron policemen, with an additional two hundred held in reserve.[60] Some four to five thousand coal and iron police were said to be in the anthracite region during the 1902 strike.[61] In 1902 alone, the Governor granted 4,512 commissions, up from 570 in 1901.[62]

With collieries and mine patches being relatively isolated and spread out among the rugged Appalachians, and with road transportation being difficult, quick deployment of coal and iron police to trouble spots was a serious problem. The Philadelphia and Reading and the Lehigh Valley companies solved this problem in a rather unique fashion. They created special coal and iron police trains which would be deployed to trouble spots. In 1888, the Reading's train consisted of two cars manned by twenty-two men. Even the cook was issued a commission and a rifle. The engine would run all the time during the winter months to provide the cars with steam heat. This experiment in police mobility apparently proved to be a success, as testimony during the 1902 Anthracite Mine Strike Commission hearings indicates that these "flying squadrons" were still in existence.[63]

As for deportment, some reports are found to commend the coal and iron police for their "Christian forebearance" under difficult circumstances, such as putting up with abusive women during the labor troubles of 1888.[64] However, other reports indicate that abuse of power by these forces was widespread. Heavy-handed tactics and questionable or unnecessary shootings and beatings were common, according to newspaper and commission reports. The list of abuses runs the gamut of police abuses which have occurred throughout American history. An ethnic dimension was present in many cases, which helped exacerbate tensions between the various ethnic groups in the region. Nevertheless, most of the violence involving coal and iron police during strikes can be better explained by their structural position and duties in strikes rather than by the individual characteristics of the policemen themselves. Had public police been in control, acts of abuse likely would have been fewer. While all police were seen by strikers as the enemy, the presence of outsiders and the direct employment of the police by the companies increased hatred of them and, no doubt, provocation against them.

In addition, most strike violence involving coal and iron police occurred as a result of their protection of strikebreakers. This was part of their normal, expected duty and, understandably, carrying it out did much to provoke union miners. There is truth to the argument of the Socialist mayor of Schenectady, New York, who said in 1917 that rather than improving police forces, labor peace would better be served by keeping strikebreakers away from scenes of strikes.[65]

In addition to protecting strikebreakers and company property, coal and iron police often were used in shows of force in order to intimidate strikers. The obvious presence of well-armed company police was itself a major irritant, sparking incidents that otherwise would not have occurred and creating what modern parlance refers to as "police riots."[66]

Indeed, the role of coal and iron police in strike riots is very similar to the role of public police in race riots.[67] However, it is complicated somewhat by the partly clandestine nature of their presence in many riot situations. Here it is necessary to understand the role county sheriffs typically played during times of strike riots.

In situations of escalating disorder, it was the sheriff's official lot to organize the restoration of law and order. The normal means of attempting to do so was through commissioning and arming special deputies to patrol the area and break up demonstrations.

This process carried with it the cloak of public legitimacy, but commonly, the sheriff did not act independently of the coal companies or their police. Often the sheriff consulted with coal company officials, used a coal company arsenal, and deputized large numbers of coal and iron police. One example of this process occurred during the famous Lattimer massacre. In this instance, nineteen strikers died and at least thirty-nine people were wounded when the heavily-armed deputies, most of whom were coal and iron police, fired into a crowd of marchers.[68]

However, as was mentioned above, cooperation with sheriffs and other local law enforcement agencies was not always forthcoming. Local authorities were caught in a serious bind during strikes. On the one hand, they were charged with maintaining public order and enforcing the law, which often required finding, arming, and paying deputies who would not be in sympathy with striking miners. In addition, the companies were prepared to offer their employees arsenals and money toward the cause, as well as exert considerable political pressure in order to get their way. Indeed, some companies would force all managerial and clerical personnel to receive coal and iron police commissions and/or act as deputies during strikes in order to retain their jobs.[69]

On the other hand, local officials were often sympathetic to the strikers and, being popularly elected, were loathe to align themselves too squarely on the side of the companies. This led to frequent conflicts between local law enforcement personnel and coal and iron police. For example, the Chief Burgess of Shenandoah blamed the 1888 riot in that city entirely on the coal and iron police, which, he said, should confine themselves to company property.[70] In another incident, the Chief of Police in Moosic was said to have "abused and threatened" the coal and iron police during the 1902 strike.[71] Those police were deemed "essential" because the sheriffs and deputies sided with the "men."

The conflict and instability among law enforcement personnel led to some interesting, and sometimes tragic, peace-keeping experiments. During the 1877 strike, the city of Scranton employed a ten-man police force, which by its small number was inadequate to keep the peace. Moreover, the force and Mayor Robert H. McKune were seen as being sympathetic to the strikers. Nevertheless, the Mayor attempted to raise a special police force to deal with the situation but was blocked by City Council. Some prominent citizens then took matters into their own hands, forming a Citizens' Corps to keep the peace. As it turned out, the Corps was anything but a neutral peace-keeping body. The group was armed by the Lackawanna Iron and Coal Company, and thirty of its forty-six members were coal and iron policemen. This Citizens' Corps was prepared to operate as a vigilante group if necessary, but the Mayor granted their request to become special policemen.

When a group of strikers marching to enlist new recruits became rowdy, the

Citizens' Corps met them and fired into the crowd, killing six and wounding fifty-four. Several indictments against Corps members were issued, but no convictions were obtained. Rather than disband, the Corps gained permanent legitimacy by becoming a unit in the Pennsylvania National Guard.[72]

By the end of the nineteenth century, then, it was clear to the miners that they were outgunned in direct confrontations with the companies. Coal operators were willing and able to use para-military force during labor disputes. When the strength of coal and iron police and local deputies was inadequate, the State would intervene with militia and the National Guard. When miners were killed or injured in battle, there was little chance that anyone would be brought to justice. In the short run, the direct military domination by the operators was effective. But over the long term, it increased bitter hatred of the police and companies by the miners. In addition, once it became clear that violent protest would not succeed, the miners were encouraged to work through labor unions, ultimately strengthening their ability to resist operator domination, especially in the anthracite fields.

The Post-1902 Era

Efforts to eliminate the coal and iron police began almost from the time they were created. Scepticism greeted their birth in anthracite country,[73] and by 1888, serious efforts to do away with them were underway.[74] However, these efforts to abolish the "private armies" had no significant effect until after the 1902 strike. The Anthracite Mine Strike Commission strongly recommended the abolition of coal and iron police and the creation of public police forces sufficient to enforce laws and keep the peace.[75] It had become clear to many that the presence of coal and iron police was an intense provocation during strikes and that a more efficient means of social control must be found. In response to the growing sentiment, Pennsylvania's Governor Pennypacker created the first State police force in the nation in 1905.[76]

The bill creating the Pennsylvania State Police reads in part that the police "are intended, as far as possible, to take the place of the police now appointed at the request of various corporations."[77] As it turned out, the State Police were more of a supplement to coal and iron police than a replacement. Legislative efforts to abolish coal and iron police upon creation of the State constabulary failed, and labor, fearing that the new force would merely be a better trained anti-union para-military organization, forced the limiting of the maximum number of State Police to 228. Clearly, this was an insufficient number to replace the coal and iron police.

Nevertheless, the role of the coal and iron police in the anthracite region diminished significantly after 1902. The settlement of that year's strike was seen to be a victory for the miners and included tacit recognition of their right to organize. In anthracite counties, United Mine Worker organization proceeded quickly. Once the area was completely organized, most strikes would shut down mines in the entire region. No one would attempt to replace the striking miners, and so picketing and disorder were minimized, as was the possibility for violent confrontations with coal and iron police.[78]

References to coal and iron police disappear almost entirely from newspapers after 1902. From other sources, it seems that their duties primarily revolved around guarding company property and guarding payrolls.[79] State Police strike reports make little reference to coal and iron police involvement. Typical is the case of a strike at a Mahanoy City stripping in 1933. It was reported that a number of men attempted to rush a coal shovel to keep it from working, and that six coal and iron policemen stationed at the site made no attempt to block the rush.[80]

Occasional shows of force were still made, although less successfully than in earlier days. An eyewitness recalls a visit of John L. Lewis to Wilkes-Barre in the 1920s, during which he spoke at the public square. Coal and iron police stood on the curbs ringing the square but did not enter the square itself. Ostensibly, this was because if they had come into the square, Lewis said he would refuse to speak, and the crowd would become hard to control.[81]

While relatively benign activities characterized the coal and iron police in the anthracite region after 1902, their significance in the bituminous region increased greatly during the 1920s, and the problems found in eastern Pennsylvania years earlier were repeated in the west.[82] Several investigations recommended abolition of these constabularies, but labor and its allies lacked the political clout necessary to end them.

However, on 10 February 1929, miner John Barkowski of Tyre was beaten to death by three coal and iron policemen.[83] This led to the passage of the Industrial Police Act of 1929, which, by the time it reached the Governor's desk, had suffered amendments which made its reforms minimal.[84]

Nearly everyone seemed unhappy with the solution, and Governor Gifford Pinchot returned to office in 1931 promising to abolish the coal and iron police. His solution was to advocate creation of a special brand of State Police which would be available to corporations *and unions* as needed, with the hiring organizations reimbursing the State for their use. In addition, the bill protected the State from any liability for wrongdoing by this special police force. This remarkable idea would have put the state in the role of trainer and supplier of mercenary troops to both corporations and unions. Only the State Police favored the bill, which was soundly defeated.[85]

Other efforts at legislative remedies also failed. Finally, on June 30, Governor Pinchot revoked all outstanding commissions and, in effect, disbanded the existing coal and iron police.[86] Even so, the law creating them was still on the books, and Pinchot moved for its removal, stating: "We have endured long enough the relics of a more barbarous age. The coal and iron police and the company-paid deputies ought to rest with the dinosaur and the great auk before the General Assembly adjourns next spring."[87] But still, political forces were inadequate.

Finally, in 1935, George H. Earle III became the first Democratic governor of Pennsylvania in the twentieth century. Democrats controlled the General Assembly and gained strength in the Senate. On June 15, the Governor signed Act 156, P. L. 348, repealing all enactments which had authorized coal and iron or industrial police.[88] Finally, sixty-nine years after their creation, the coal and iron police were abolished.

In effect, all of this made little difference in anthracite country in terms of the function of company police, which had been primarily that of watchman since 1902. Organizationally, companies either hired watchmen or contracted with private security firms to supply them. In the case of the Reading Company, the latter course was taken, although the personnel of the old coal and iron police force remained as security guards under the new arrangement, and were still paid directly by the company and quartered on company property.[89] But with their State power gone, and with the continued decline of the anthracite industry, their numbers and importance diminished. However, their tradition lives on to this day, Reading watchmen are often referred to as coal and iron police.

Significance

Pennsylvania's coal and iron police were one of a number of quasi-public police organizations spawned by the economic and social conditions of the mid-to-late nineteenth century. In the anthracite region as in many others, immigrant groups which had been used to the pace, life style, and work rhythms of a rural agricultural society were thrust into industrializing America. As Gutman has pointed out, adjustment was difficult, and employers faced severe problems in convincing each new immigrant wave to accept its lot of long hours of brutal work without job autonomy or security.[90] This was particularly hard when wages fell, which happened often during the late nineteenth century because of frequent recessions and the instability of the coal industry as a whole. Social and labor unrest were inevitable outcomes of such conditions.

Business, then, had to find ways of establishing social control. The reliability of local law-enforcement agencies was uncertain at best, due to the ability of miners and their supporters to gain and sustain sometimes significant political power at the local level.[91] Therefore, an extra local public force or a force controlled directly by business were the alternatives to be considered. The former solution proved to be politically impractical until the twentieth century. Therefore, the latter was chosen.

In the short run, these forces did what was required. They aided their employers in maintaining corporate control over the work force and protected their employers' property and property rights. They patrolled areas of the coal regions that local officials could not cover adequately, and while cloaked with State authority, were directly answerable to the corporations whose interests they represented. Their use was able to lessen the number of times State militia or National Guard troops were needed, saving the State much money. They could be expanded quickly during labor disputes, then disbanded just as fast when trouble subsided. Their presence left no doubt as to who controlled the coal regions.

However, in the long run, the coal and iron police proved to be more detrimental than useful to the corporations' interests. As Johnson states, concerning business use of private police generally in the nineteenth century, "This development profoundly undermined the American legal order and the public order generally."[92] This fact did not go unnoticed and hit home particularly hard to those who came under the coal and iron police's jurisdiction. These forces provided a common focus for hatred by miners, being perhaps the most visible symbol of their oppression. Where coal and iron police were present, direct domination by largely absentee coal companies was starkly visible.

In the anthracite region of the late nineteenth century, a kind of internal colonialism existed whereby absentee coal companies exploited the region and its people.[93] Coal and profits flowed out of the region, and little was given back to it. As with regular colonialism, this internal colonialism was not a very good system of domination in the long run.[94] It was too direct, power was blatant and unmasked. Direct colonialism has become less evident over time, being replaced by other more subtle forms which make it easier for, in Gramsci's terms, "the dominant fundamental group" to secure "spontaneous consent" from the oppressed.[95] Spontaneous consent is very difficult to establish when the subject group is policed by direct employees of the dominant group. Alternatively, a public professionalized force is less open to criticism for being biased or irresponsible. If the public force is well trained, instances of gross violent abuse will be fewer. If the force, and the public, is convinced that the public police are representing the best interests of society as a whole, as opposed to any particular group or class, the volatility of police presence is neutralized significantly. Hence, the creation of the Pennsylvania State Police (the first state police force in the country) and the eventual demise of the coal and iron constabularies.

Seen from this angle, the coal and iron police were used during a time when the State was unable to adequately protect dominant class interests. But by

1902, problems with this solution were apparent, and by 1935, armed with a successful State Police force, the State was prepared to assume its full responsibility of providing professionally trained police to guarantee the stability of class and property relations. The direct internal colonial system of domination was changing to a "neo-colonial" system, one less transparent and onerous, but, in the long run, more stable and effective.

NOTES

[1] For overviews of the history of police and their functions, see Abraham S. Blumberg and Arthur Niederhoffer, "The Police in Social and Historical Perspective," in Abraham S. Blumberg and Arthur Niederhoffer (eds.), *The Ambivalent Force: Perspectives on Police* (Waltham, Mass., 1970), 1–15; Gary T. Marx, "Civil Disorder and the Agents of Social Control," *Journal of Social Issues,* 26 (1970), 19–57; Bruce Smith, Sr., *Police Systems in the United States,* 2nd rev. ed. (New York, 1960); and Samuel Walker, "The Urban Police in American History: A Review of the Literature," *Journal of Police Science and Administration,* 4 (1976), 252–260.

[2] For accounts of the early history of coal in the United States, see Frederick Moore Binder, *Coal Age Empire* (Harrisburg, 1974); and Howard N. Eavenson, *The First Century and a Quarter of the American Coal Industry* (Pittsburgh, 1942).

[3] Ben J. Wattenberg (ed.), *The Statistical History of the United States* (New York, 1976) 590, 592.

[4] Clifton K. Yearly, Jr., *Enterprise in Anthracite* (Baltimore, 1961), 30–31, 83–84.

[5] Marvin W. Schlegel, *Ruler of the Reading: The Life of Franklin B. Gowen, 1836–1889* (Harrisburg, 1947), 5; Harold W. Aurand, *From the Molly Maguires to the United Mine Workers: The Social Ecology of an Industrial Union, 1869–1897* (Philadelphia, 1971), 142.

[6] Yearley, *Enterprise,* 209, 212.

[7] John K. Mumford, *Anthracite* (New York, 1925), 81.

[8] *Proceedings of the Anthracite Mine Strike Commission* (Scranton, 1902–1903), 274. On the history of cooperation in price and market control between various railroad and coal companies, see Eliot Jones, *The Anthracite Coal Combination in the United States* (Cambridge, 1914); and Schlegel, *op. cit.*

[9] E. Digby Baltzell, *Philadelphia Gentlemen: The Making of a National Upper Class* (Glencoe, Ill., 1958), 118.

[10] Aurand, *From the Molly Maguires,* 20; Muriel Earley Sheppard, *Cloud By Day: The Story of Coal and Coke and People* (Chapel Hill, 1947), 109–110.

[11] Jeremiah Patrick Shalloo, *Private Police: With Special Reference to Pennsylvania* (Philadelphia, 1933), 103; Sheppard, *Cloud By Day,* 113–119.

[12] Aurand, *From the Molly Maguires,* 21; Sheppard, *Cloud By Day,* 109–110.

[13] Aurand, *From the Molly Maguires,* 21–22.

[14] *Ibid.,* 26, 22.

[15] Yearly, *Enterprise,* 206–207.

[16] Aurand, *From the Molly Maguires,* 25. See also Wayne G. Broehl, Jr., *The Molly Maguires* (Cambridge, 1964), 295–296.

[17] On the concept of cultural division of labor, see Michael Hechter, *Internal Colonialism: The Celtic Fringe in British National Development, 1536–1966* (Berkeley, 1975).

[18]Broehl, *Molly Maguires,* 79–85; William A. Gudelunas and William G. Shade, *Before the Molly Maguires; The Emergence of the Ethno-Religious Factor in the Politics of the Lower Anthracite Region,* (New York, 1976); Aurand, *op. cit.*, 27–29. Ethnic divisions are still surprisingly strong today within certain areas of Pennsylvania coal country; see William Gudelunas and Stephen R. Couch, "Would a Protestant or Polish Kennedy Have Won? A Local Test of Ethnicity and Religion in the 1960 Presidential Election," *Ethnic Groups: An International Periodical of Ethnic Studies,* 3 (December 1980), 1–21.

[19]Rowland Berthoff, "The Social Order of the Anthracite Region, 1825–1902," *Pennsylvania Magazine of History and Biography,* (July 1965), 261–291.

[20]*Legislative Record* (Harrisburg, 1864), 970.

[21]*Legislative Record* (1866), 459–460.

[22]*Legislative Record* (1867), CCXXII.

[23]*Ibid.,* CCXXIV.

[24]*Laws of Pennsylvania* (1865), 225.

[25]*Laws of Pennsylvania* (1866), 99.

[26]Shalloo, *Private Police,* 61; Sheppard, *Cloud by Day,* 19–20.

[27]Shalloo, *Private Police,* 61.

[28]James Q. Wilson, "What Makes a Better Policeman?" *Atlantic Monthly,* 223 (1969), 129–135; Gresham M. Sykes, *Criminology* (New York, 1978), 368–369; Steven Spitzer and Andrew T. Scull, "Privatization and Capitalist Development: The case of the Private Police," *Social Problems* (October 1977), 21–22; Bruce C. Johnson, "Taking Care of Labor: The Police in American Politics," *Theory and Society,* 3 (Spring 1976), 94.

[29]Smith, *Police Systems,* 66–84.

[30]Shalloo, *Private Police,* 82.

[31]Sheppard, *Cloud by Day,* 110; Commission on Special Policing in Industry, *Report to Governor Gifford Pinchot,* Special Bulletin No. 38 (Harrisburg, 1934), 16.

[32]Mary Siegel Tyson, *The Miners* (Pine Grove, 1977), 30–31.

[33]*Publications of the Historical Society of Schuylkill County* (Pottsville, 1907), 337.

[34]Adolf W. Schalek and D. C. Henning (eds.), *History of Schuylkill County, Pa.* (Harrisburg, 1907), vol. 1, 158.

[35]*History of Schuylkill County* (New York, 1881), 53–54.

[36]Eavenson, *American Coal Industry,* 378.

[37]Broehl, *Molly Maguires,* 86–101.

[38]Shalloo, *Private Police,* 59; Sheppard, *Cloud by Day,* 111; Aurand, *From the Molly Maguires,* 91; Broehl, *Molly Maguires,* 203–209.

[39]The number of yearly commissions granted is on record. However, these figures do not give an accurate idea of the number of active coal and iron policemen. Throughout most of their history, there was no limit to the length of coal and iron police commissions, so the number on duty from previous years was not known. In addition, companies would discharge coal and iron policemen without notifying the State.

[40]Coal and Iron Police Commissions, Appointments File (Geary Administration), Department of State (Commissions), Division of Archives and Manuscripts, Pennsylvania Historical and Museum Commission.

[41]Wilson, "A Better Policeman," 130.

[42]Shalloo, *Private Police,* 127.

[43]*Ibid.,* 110–111.

[44]Perry Stirling, *Bootleggers, Breakers and Beer* (Summit Hill, Pa.: American Printing Company, 1974), 7.

⁴⁵Sheppard, *Cloud by Day,* 98–99; Shalloo, *Private Police,* 111.
⁴⁶Aurand, *From the Molly Maguires,* 25, 138–141; Michael Novak, *The Guns of Lattimer* (New York: Basic Books, 1978).
⁴⁷Shalloo, *Private Police,* 110–111.
⁴⁸Sheppard, *Cloud by Day,* 111.
⁴⁹Aurand, *From the Molly Maguires,* 25.
⁵⁰Primary sources on the Mollies are scarce and the complete story may never be known; see Harold W. Aurand and William Gudelunas, "The Mythical Quality of Molly Maguire," *Pennsylvania History,* 49 (April 1982), 91–105. Still, many have attempted to piece together what is available. The best, most balanced treatment so far is Broehl, *Molly Maguires.* Other works of note are Anthony Bimba, *The Molly Maguires* (New York, 1932); Tom Barrett, *The Mollies Were Men* (New York, 1932); and Arthur H. Lewis, *Lament for the Molly Maguires* (New York, 1964).
⁵¹Anna Rochester, *Labor and Coal* (New York, 1931), 165.
⁵²Broehl, *Molly Maguires,* 131–151.
⁵³George R. Leighton, *Five Cities* (New York, 1939), 9.
⁵⁴*Daily Miners' Journal,* 8 April 1975.
⁵⁵*Ibid.,* 7 September 1897.
⁵⁶*Shenandoah Evening Herald,* 31 July 1877.
⁵⁷*Pottsville Republican,* 4 February 1888.
⁵⁸*Proceedings of the Anthracite Mine Strike Commission,* 196.
⁵⁹*Pottsville Republican,* 3 January 1888.
⁶⁰*Ibid.,* 7 February 1888.
⁶¹*Ibid.,* 31 May 1902; Shalloo, *Private Police,* 85.
⁶²Philip M. Conti, *The Pennsylvania State Police* (Harrisburg, 1977), 35.
⁶³Testimony, U.S. Anthracite Mine Strike Commission (1902–1902), 7323; see also Louis Poliniak, "The Coal and Iron Police," *Citizen Shopper,* 12 June 1974.
⁶⁴*Daily Miners' Journal,* 16 February 1888.
⁶⁵Conti, *State Police,* 124.
⁶⁶*Proceedings of the Anthracite Mine Strike Commission,* 293–293; Commission on Special Policing in Industry, 11.
⁶⁷Marx, "Civil Disorder," 21; Walker, "The Urban Police," 138–139.
⁶⁸Novak, *Lattimer Massacre;* Aurand, *From the Molly Maguires,* 138–139.
⁶⁹Testimony, U.S. Anthracite Mine Strike Commission (1902–1903), Mr. James H. Lenahan.
⁷⁰*Pottsville Republican,* 4 February 1888.
⁷¹Testimony, U.S. Anthracite Mine Strike Commission (1902–1903), 9413.
⁷²Samuel C. Logan, *A City's Danger and Defense* (Philadelphia, 1887); Aurand, *From the Molly Maguires,* 111–114.
⁷³*Miner's Journal,* 14 April 1866.
⁷⁴*Daily Miners' Journal,* 25 February 1888.
⁷⁵*Proceedings of the Anthracite Mine Strike Commission,* 293–294.
⁷⁶Conti, *State Police,* 35–40.
⁷⁷*Ibid.,* 40.
⁷⁸Commission on Special Policing in Industry, 14-15.
⁷⁹Interviews with William Jeffries and Anthony Pickard, retired employees of Reading Anthracite Company, Pottsville, Pa., 12 July 1982.
⁸⁰Memo from Reese L. Davis to Commanding Officer, Troop "C", State Police, 18 September 1933, from Strike Reports, Pennsylvania State Police, Division of Archives and Manuscripts, Pennsylvania Historical and Museum Commission.

[81]Interview with Edward Fox, retired President, Reading Anthracite Company, Pottsville, Pa., 7 July 1982.
[82]Shalloo, *Private Police,* 59; Commission on Special Policing in Industry, 16; Sheppard, *Cloud by Day,* 107–108.
[83]Frank Butler and Robert Taylor, "Coal and Iron Justice," *The Nation,* 129 (16 October 1929), 404–405; "Good Men and True: The Story of the Barkoski Trial," *New Republic,* 60 (30 October 1929), 292–293; Shalloo, *Private Police,* 65–72.
[84]*Legislative Journal* (Harrisburg, 1929), vol. 2, 2420–2422, 2624–2625, 3097–3098, 3229–3230; vol. 3, 4762; *Laws of Pennsylvania* (1929), 546–549.
[85]*Legislative Journal* (1931), vol. 2:2612–2614, 3054–3063.
[86]Commission on Special Policing in Industry, 19–20; Shalloo, *Private Police,* 81.
[87]Commission on Special Policing in Industry, 5.
[88]*Legislative Journal* (1935), vol. 1, 358–359; *Laws of Pennsylvania* (1935), 348–349.
[89]Interview with Edward Fox, *op. cit.*
[90]Herbert G. Gutman, "Work, Culture, and Society in Industrializing America, 1815–1919," in *Work, Culture and Society in Industrializing America* (New York, 1977), 3–78.
[91]Spitzer and Scull, "Privatization and Capitalist Development," 22.
[92]Johnson, "Taking Care of Labor," 95.
[93]On the concept of internal colonialism, see Hechter, *Internal Colonialism;* Pablo Gonzales-Casanova, "Internal Colonialism and National Development;" *Studies in Comparative International Development,* 1 (1965), 27–37; Paul Nyden, "An Internal Colony: Labor Conflict and Capitalism in Appalachian Coal," *Insurgent Sociologist,* 8 (1979), 33–43; Stephen R. Couch, "Internal Colonialism: A Re-examination," paper presented at the Pennsylvania Sociological Society Annual Meeting, University Park, Pa (15 November 1980); and Stephen R. Couch, "Social Control in an Internal Colony: Pennsylvania's Coal and Iron Police," paper presented at the American Sociological Association Annual Meetings, Toronto (24–28 August 1981).
[94]See, for instance, Immanuel Wallerstein, "The Rise and Future Demise of the World Capitalist System: Concepts for Comparative Studies Analysis," *Comparative Studies in Society and History,* 16 (1974), 387–415.
[95]Antonio Gramsci, *Selections from the Prison Notebooks* (New York, 1971), 12.

*I am grateful to Lisa David, Joyce Janowski, Marie Kahler, Linda Myers Lazar, Rosanne Troy, and the Schuylkill County Arts and Ethnic Center for their valuable assistance on this paper. The research was partially supported by grants from the Faculty Scholarship Support Fund an the College of Liberal Arts of the Pennsylvania State University.

Chapter Nine

COMMENTARY: THE FAMILY ECONOMY AND LABOR PROTEST IN INDUSTRIAL AMERICA AND THE COAL AND IRON POLICE IN ANTHRACITE COUNTRY

Ronald M. Benson

The two presentations we have just listened to, by Stephen Couch and John Bodnar, are based on obvious and substantial topics which relate to the essential life-scape of Pennsylvania anthracite communities during the last one hundred years. The papers on which these presentations were based indicate careful research, a sustained consideration of evidence drawn from primary and secondary sources, both print and oral, and a commendable concern to plumb the deeper, underlying meanings of the social phenomena of late nineteenth-century quasi-legal social control in the coal fields and the varied communal responses of miners and their families to the Depression of the 1930s. Both authors attempt to interpret and order their evidence with attention to economic or psychological theories drawn from the broader social sciences. How well they have tested those theories within the context of their evidence is now the subject of discussion. As Sir Ivor Noel Hume, in his recent study of *Martin's Hundred,* reminds us, as well as himself, the most attractive theories are often undone by the eroded nail and the silver eyelet.

First, I wish to take up the paper by Professor Bodnar, or should I say "papers," for clearly he has written two separate studies here. One paper focuses on the split between the United Mine Workers (UMW) and the United Anthracite Miners (UAM), and the second attempts to explore the psychological impact of the Great Depression on first and second-generation miners and their families as they responded to unemployment and intra-familial tensions growing out of sustained pressure on the "family economy" by the downward thrust of the business cycle in the 1930s.

I wish to deal with each of Bodnar's papers separately, but the second in a more critical way then the first. I agree with Bodnar that it is obviously shortsighted to interpret the life of the American worker from the perspective of the shopfloor, and the shopfloor only, as though life were nothing more than the extension of a shovel handle or a conveyor belt. Every work day comes to an end. Labor historians still seem bent on isolating the American worker from society as a whole, to ignore the parameters of the business culture, to concentrate on the romanticism of militancy, or to assume that labor history is merely the choreographed progression of individuals across a simple stage. It seems to me that these points have all been made before, whether or not we continue to write labor history within an institutional framework. American workers were essentially *conservative;* family was often more important than the job, religious values more important than secular dogma, and social linkages more attractive than political ideology.

Bodnar explains the emergence of the UAM as a "conservative" community-based movement, acting in revulsion to the Lewis-dominated UMWA, whose solution after 1930, was to urge that the work in the mines be shared. The UAM movement was thus reformist, seeking to restructure not only the order of the workplace but also that of the union which exercised joint control of it. "Job equalization" reflected a deep-seated impulse to share the available work in order to preserve the community. This was, of course, another form of "soldiering," and one long denounced by F. W. Taylor for its conservative resistance to change. John L. Lewis refused to buy "job equalization" in the end because it threatened the rationalized *modus vivendi* established between the coal operators and the union for stabilization of the workplace. As any intelligent union member knows, collective bargaining creates two "bosses," where only one has existed previously. The "checkoff" of union dues made it easy to detach the union from the democratic control of its members, thereby entrenching the trade-union bureaucracy. The momentary revulsion against Lewis in the anthracite fields in the early 1930s was neither the first revolt nor the last. In the end, however, Lewis won. When prosperity returned briefly during World War II, it was easy enough to forget the revocation of charters, provisional control of union locals, and the "pie cards" who staffed the national and local offices. Controlled collective bargaining became little more than the playing out of a ritual struggle between capital and labor. As Bodnar notes, the "job-equalization" movement was a direct manifestation of communal solidarity, and at its most basic level was unsettling but conservative; that it appeared "radical" was the result of the collusion between operators and union bureaucrats, who saw communal action as a disruptive challenge to the stability of their joint control of the unionized sector of the coal industry.

The second "paper," linked to the first by the theme of collective or communal responses to the major tensions of the 1930s, covers pages 89–95. Here Bodnar attempts to examine the impact of unemployment and the Depression on familial and communal solidarity in several anthracite communities. There was, he tells us, a crucial "family economy" which bound parents to children and children to parents. In particular, he suggests that there was a casual relationship between the Depression, measured no doubt by diminished employment and income, although this is not quite made clear, and an alarming increase in psychotic illnesses, primarily schizophrenia and manic depression. Drawing from a number of interviews conducted with residents of anthracite communities in a period from 1977 to 1981, Bodnar is able to reconstruct a picture of strong familial ties, early employment experiences, discipline patterns, and family strategies for survival. While inherently interesting, do the data in fact support the application of widespread dementia praecox diagnosis to a generation of American-born males, primarily Eastern or Central European, of crucial age and in a depressed economy? I will not belabor the point of whether dementia praecox is the result of a crucial chemical imbalance, a distorted mother-son relationship, or some combination of the two. It is beyond my competence to do so. There are, however, a number of questions I wish to raise about this approach, the interpretation of the data, and what it may *not* tell us about life in the anthracite settlements in the 1930s. Were these communities really any different from communities in a similar setting with similar population size or mix?

First, Bodnar does not always define such crucial concepts as "family economy." In fact the family economy he sketches out existed beyond the coal fields of the anthracite region, existed in fact before the rise of capitalism, and in fact exists to this day. What was unique about the family economy as observed by Central or Eastern Europeans living in the region? Was their family economy any different from that of the Irish or other ethnic groups living in those communities? Second, Bodnar fails to measure or define the Depression in economic terms which are concrete and lend themselves to the analysis suggested by his model. We can all admit, I think, that economic hard times quickly produce emotional tensions within the family. In the 1930s, despondency, helplessness, and frustration became hallmarks of the Depression, but so did mobility. What was the unemployment rate; what was the average workweek, for those lucky enough to work; what was their average annual income; and how many workers actually saw a rise in real wages as prices dropped faster than income? When these questions are answered, we can measure in a more precise way the stress on the family economy. Third, one must approach with a great deal of caution the relationship between class and medical diagnoses. The lower classes, as we now suspect, tend to be over-diagnosed, over-medicated (and this is particularly true for dementia patients), and over-institutionalized. Admission records do not constitute clinical evidence. Fourth, the sample used by Bodnar may be weighted in favor of the working-class population of the area and therefore tells us little when weighted against a national sample not so weighted. Fifth, the Depression of the 1930s might not have led to an alarming intensification of dementia praecox, but a greater impulse to institutionalize (perhaps among his sample in particular), since unemployment and diminished income made it harder to care for those suffering *some* symptoms of the disease as wage-earners lost jobs and members of the family moved back home. Finally, one might well question the competency of admitting personnel to make an accurate diagnosis, since dementia has often been used for nothing more than a categorical catch-all for record keeping purposes.

Finally, I wish to suggest that Bodnar did not put the right questions to those interviewed, or at least did not use such data in this paper, in order to pursue this question. It is unlikely that many of those interviewed would talk freely about such "family" problems, but let me give several examples of the types of questions that might be pursued in this case. Two manifestations of schizophrenia are autistic withdrawal and severe ambivalence. One can certainly follow-up questions dealing with the behavior patterns of sons in the early to late 1930s. Was the typical response to withdraw, to seek refuge in one's room, to sleep a good part of the day to escape the tensions and anxiety of the household? What might police records show about visits made to quell home disturbances in this period? Did such police calls increase in this period? How ambivalent were those second-generation males to the challenges raised by the "job-equalization" movement and the challenge to the UMWA? Who supported the UAM? As two respected psychiatrists have noted: "One of the most difficult problems of research in schizophrenia has been the fact that this diagnosis has become a waste basket." In this case, Bodnar's data simply will not support his contention.

On the other hand, the suggestion that the Great Depression led to a rise in manic depression among foreign-born females who played dominant roles in the family is easier to align with the evidence. Normal tensions of adjusting to a new culture, coupled with the economic collapse of the 1930s, easily increased the incidence of manic depression, which is easier to diagnose. However, what is not so easily derived from Bodnar's data is a clear picture of the relationships of mothers and sons. In both cases, this may really be the critical issue.

Now, may I turn to the paper on the coal and iron police by Professor Couch? This study also invokes the sanction of theory to order the evidence. Couch argues that the anthracite region existed as an "internal colony" where absentee coal companies exploited the region and its people. Coal and profits flowed out of the region, and little was given back to it." While this is currently very good neo-Marxist rhetoric, what does such a blanket assertion really tell us about the sociology of the region and the emergence of the coal and iron police as a quasi-legal force to oversee order and stability in the anthracite region. Let us assume that the law exists to protect and defend property rights, whether those rights be exercised by individuals or embodied in the functions of the State. In the period after 1866, the coal and iron companies demanded protection of their property rights which the state could not provide, or at least was unwilling to provide within the context of privatism which continued into the nineteenth century. Where costs could be borne by private individuals or companies, or local governmental units, there was no reason for the state to assume those costs. This was the pattern in social services from garbage collection, to sanitary drinking water, to public recreation facilities. In part the development of the coal and iron police can be explained by the continuation of eighteenth century "privatism" into the nineteenth century and the general reluctance of community leaders to abandon that ideology in favor of a more direct intervention by the state. In this case, the coal and iron companies demanded the kind of protection of their property rights which the state could not provide, and in the vacuum thus created the companies moved to establish their own devices for order and stability. Agents for law and order at the local level were not viable instruments of order; the collapse of local and county government, democraticly-based, ushered in the era of quasi-legal private police.

Violence, moreover, was no stranger to the coal fields even before the appearance of the private police. Workers attacked fellow workers, vented their frustrations against the property of others, and resorted to collective action to defend group interests. Patterns of violence were legitimized by cultural experiences prior to immigration, or by frustrations engendered by the discipline imposed by industrialization.

There are a number of questions in regard to the rise of the coal and iron police which still require answers, despite Couch's efforts to re-examine their "legitimacy" within a framework of ineffective state or local government or industrialization of still rural areas. What we would still like to know, for example, is who were those recruited to serve in the iron and coal police? Were the English and Welsh recruited to discipline the Irish? By the 1880s, were the Irish recruited to discipline Central and Eastern Europeans? Did ethnic tensions serve as the major recruiting appeal for the private police? Were many of the coal and iron police recruited within the communities in which they served? Who was recruited from Philadelphia or Pittsburgh or Camden, so that local or familial ties would not interfere with the assignment of duties? How long did a typical coal and iron policeman serve? Did slack times in the mines produce numerous local recruits, and flush times furnish few? Did those recruited serve their time and then settle in those communities as storekeepers, saloon owners, or company officials? Such questions are suggested but not quite answered by Couch's paper.

Couch's suggestion that the presence of the coal and iron police in the anthracite area eventually amounted to military occupation of the region hardly seems borne out by the evidence. The judicial system did not collapse, to be replaced by martial law, even though justice was a sometime thing. After

1920, the coal and iron police became a political liability which politicians were eager to shed. It is true that the creation of the State Police merely shifted in many cases the quasi-legal functions of the private police to the new State Police. By the 1920s, however, the new State Police seemed professional enough to handle strikes and labor disturbances in a more competent manner; in truth, state and local police were at times co-opted by the companies to patrol the towns, pick up suspected union organizers, and escort unwanted strangers away from the town limits. But they did this not only in the hard-coal region, but in the steel towns of Ohio, the textile towns of the South, and the soft-coal regions of the West. Furthermore, the quasi-legal status of the coal and iron police established a tenuous line of social accountability; in other areas of the nation where no such accountability was established, the Klan or vigilantism exercised the same functions of maintaining order and serving the interests of proprietary powers of dominant social groups. The Wobblies, Communists, and CIO organizers, not to mention pacifists or harmless gas-and-water socialists, encountered similar methods of community control in other regions of the nation. This raises the question, of course, of how the coal and iron police really provoked the strengthening of a strong union movement in northeastern Pennsylvania. One must approach this question from the trade unionist position, an approach not used in this paper.

Couch's final contention that private police functioned within the economic context of internal colonialism adds, it seems to me, little or nothing to his analysis. Colonialism was often beneficial to the colony, or colonies, as in the American experience. The hard-coal region of Pennsylvania was not marginal to the development of industrial or finance capitalism but at the core of those crucial economic phases. In a capitalist economy there is no idle capital and profits always flow somewhere, even into union pension funds. Not all owners were by any means absentee owners; the Scranton family remains a prominent family in Pennsylvania politics.

Good papers raise questions; I hope I have not suggested something to the contrary. In raising some of these questions, or in questioning some of the conclusions of Professors Bodnar and Couch, I am simply complimenting two stimulating researchers and scholars. I have also reminded myself how much easier it is to critique a paper than it is to write one.

Chapter 10

ETHNIC RESPONSES TO THE LATTIMER MASSACRE

George A. Turner

In the small, coal-mining patch of Lattimer, a company town owned by Pardee Coal Company a few miles northeast of Hazleton, Pennsylvania, one of the worst tragedies in American labor history occurred on Friday, September 10, 1898. It happened as a consequence of a strike started by immigrant coal miners in the Hazleton region who were seeking higher wages and a redress of their grievances. These laborers felt oppressed and exploited and were determined to rectify their plight. In an effort to expand the strike a group of nearly four hundred unarmed immigrant mine workers marched from Harwood to Lattimer to get their fellow workers at the Pardee mine there to join them in a work stoppage. On the outskirts of Lattimer they encountered Luzerne County Sheriff James Martin and his eighty-six deputies, who saw the strikers as enemies of public order and the wellspring of lawlessness. What evolved was a confrontation between two diametrically opposing forces, law officers and resolute strikers, which resulted in a calamity of violent proportions.

The sheriff approached the marchers and ordered them to halt, disperse, and return to their homes. His directive unheeded, an altercation quickly erupted between Martin and the leaders at the head of the marching column. At this point a number of deputies, without advance warning, raised their rifles, aimed directly at the unarmed strikers, and fired. Instantly, their gunshots produced a carnage of death and injury—a massacre that left nineteen dead and thirty-eight wounded.[1] After the deadly volley ceased and the smoke cleared, one could hear the groans and shrieks of the strikers; the mutilated and bloody bodies of the dead and injured produced stares of disbelief. Unscathed in this slaughter of defenseless workers, peacefully assembled on a public highway, were the peace officers. This tragedy, lasting only a few minutes, became the most serious act of labor violence in Pennsylvania's history and nationally one of the most devastating, in which public authorities were responsible for attacking, wounding, and killing American laborers.

This terrible event, frequently called the "Lattimer massacre," witnessed a deadly collision of two divergent groups. The English-speaking establishment was determined to stop the strikers, who were eastern European immigrants, their opponents, imbued with a spirit to win the strike. Those who were of the same heritage as the victims were not silenced by the roar of the gunshots nor made apathetical by their lack of prestige or standing in American society. This paper will explore different responses of the ethnic community to the tragedy that befell some of their own members at Lattimer.

News of the shooting immediately spread throughout the Hazleton vicinity, causing people to become alarmed, excited, and indignant over the tragedy. Uncertainty and concern prevailed in the area as to what would be the responses from the foreign community whose members were both the victims and those suffering the grievous losses. The circumstances inspired all kinds of rumors of possible disorders that could lead to riots. Talk was rampant that blood-thirsty mobs of immigrants would hunt down the sheriff and his deputies for revenge.[2] Individuals milled around in the streets trying to learn more of what happened at Lattimer; a large crowd stood outside the hospital to learn the fate of those who were injured. A pitiful sight was the numerous relatives and friends of the dead who had come to different mortuaries to identify their loved ones.

Those feeling the sting of the Lattimer bullets were members of the eastern European immigrant society. Living in a strange new country, they felt the despair of being isolated, the frustration of being powerless, and the pain of exploitation. Attracted to a nation that inspired hope, they suddenly faced a situation in which members of their own group were, in their eyes, mercilessly killed and injured by a sheriff's posse without any justification. Underscoring this sense of injustice was a report in which a mine foremen, who knew the Italian, Hungarian, and Polish workers, expressed the opinion "that if the strikers in the Hazleton region were of the English-speaking class there would have been no bloodshed."[3]

Straz, one of their newspapers, wrote: "The feelings of the masses are hurt to such a degree that it is impossible to carry on their daily routine and business the way it happens after any sensational murder."[4] It described some of the emotions of the working class in the Hazleton area as those of anger, fear, and helplessness. *Svoboda,* another Slavic newspaper in the anthracite region, expressed the viewpoint that nativist sentiments were a major factor in the shooting. "Knowing with what hatred is breathing every capricious American against any Slavonic man, who comes under general classification as 'Hungarian,' it can be said with certainty, that the sheriff ordered to shoot toward hated Hungarians at the first little resistance on the part of the workers."[5] Using a phrase from Lincoln's Gettysburg Address, *Straz* expressed a hope for those who died at Lattimer: "May their death not be in vain, may they become the patron saints of the working people in America."[6]

The events at Lattimer produced an outpouring of indignation and denunciation of the sheriff and his deputies in the Hazleton area. Rev. J. V. Moylan of St. Gabriel Roman Catholic Church described what happened at Lattimer as a "brutal and unjustifiable massacre."[7] Rev. Francis I. Pribyl, the former priest of St. Joseph's Slavonian Church in Hazleton where many of the marchers were members, observed that the men faced a desperate situation. He charged that the men worked under cruel conditions and their families faced starvation conditions. Pribyl, speaking as an advocate for the Slavic community, asserted:

The slaughtering of them in cold blood is the most high-handed piece of butchery that has ever been perpetrated upon a peaceable people and a sad commentary upon the boasted freedom of fair Columbia, whose extended arms we are taught to believe are continually outstretched to the down trodden and oppressed people of other lands.[8]

The city's mayor, Justus Altmiller, in a newspaper interview stated a similar outrage:

All I can say is that I call this shooting a butchery. I can see no excuse for the Sheriff's people having shot these men. There is no doubt in my mind that the Sheriff and the deputies lost their heads. Had they been cool, calm and collected, had they looked upon the situation with care, this slaughter would never had occurred and the name of our good city would never have been besmirched as it is today.[9]

The Hazleton newspaper, *The Daily Standard,* expressed sentiments corresponding to Altmiller's views in its headline announcing the Lattimer tragedy: "Yesterday's Butchery—A Mob of Heartless Deputies Fire Into a Throng of Marchers and Accomplish Deadly Work."[10] Other headlines in regional newspapers described the calamity: *Pottsville Republican,* "Strikers Shot in Cold Blood";[11] *Wilkes-Barre Record,* "Day of Blood at Lattimer";[12] and South Bethlehem *Daily Globe,* "Looks Like Butchery—Mob of Heartless Deputies Fire Upon Strikers."[13] The story quickly spread beyond the anthracite region. Major metropolitan newspapers depicted the shooting in equally stark terms: *The New York Tribune,* "Strikers March to Death";[14] *Detroit Free Press,* "Laid Low By Bullets, the Men Fell Like Sheep Before the Murderous Winchesters of the Officials";[15] *Boston Daily Globe,* "Dead in Heaps, Deputies Fire on Miners at Lattimer, Penn., Men Were Huddled Closely and Slaughter was Terrific,"[16] *St. Louis Post Dispatch,* "The Killing of Miners at Lattimer was Butchery";[17] *Pittsburgh Dispatch,* "Homestead Battle Thrown in the Shade";[18] and *San Francisco Chronicle,* "Mowed Down by Deputies."[19]

In the Slovak press there were outcries of indignation, calling September 10 a day of infamy. In *Amerikansko Slovenske Noviny,* its headline announced: "Massacre of Slavs—in the Freest Country Under the Sun—People are Shot at like Dogs—Slavs are the Victims of American Savagery."[20] The editor, Pucher Ciernovodsky, wrote:

> The mountainsides of Hazleton are drenched with Slovak blood, and pitiable orphans and widows, fathers and mothers, brothers and friends raise eyes brimming with bitter tears to heaven pleading, O God, is there justice in this life? If there is Justice on this earth of ours, did our blood truly deserve this spilling? Who in this world is committed to greater drudgery than we? Who in this world is engaged in a work that is more dangerous than ours?[21]

When the Slovak immigrant community learned what had happened to its fellow countrymen, feelings of outrage and consternation ignited as never before in Slovak-American life. This grief over Lattimer caused a coming together, a closing of ranks, a unity that submerged factional differences. The corporate pain caused a rallying spirit, a call to help the victims and to protest this terrible act of inhumanity at Lattimer.[22]

Many people felt that the actions of the sheriff's posse were nothing more than "official murder." This sense of indignation found expression in a number of public meetings held not only in the Hazleton area, but in the anthracite region, and in the metropolitan centers of Chicago and New York. Within two and one-half weeks the coal mining towns of eastern Pennsylvania, Shenandoah, Shamokin, Nanticoke, Mt. Carmel, Plymouth, Scranton, Wilkes-Barre, Duryea, Edwardsville, and Pittston all had meetings protesting what occurred at Lattimer.[23] These rallies basically sought to mobilize support for the victims and their families and to urge the prosecution of the law officers for murder. They were attended primarily by people of an eastern European heritage, usually with clergymen, particularly Catholic priests, as active participants, and organized either by the leadership in ethnic fraternal organizations, unions, or the Socialist party.

The start of these indignation meetings began in Hazleton, a few hours after the shooting, when some two thousand people met to express their sorrow for the Lattimer victims. The Reverend Mr. Spaulding, a Baptist minister, one of several speakers, called for those guilty of killing and injuring the unarmed strikers to be punished. Others gave assurances that assistance would be provided for the injured and the families of the dead. John Fahy, District President of the United Mine Workers of America, cautioned the audience to be calm and peaceful. Before the crowd dispersed, it unanimously adopted a series of resolutions declaring its sympathy and financial support for the Lattimer sufferers, denouncing the actions of the law officers as unwarranted, calling for their prosecution, and opposing the sending of troops into the area as many anticipated.[24]

The next day as expected, Governor Hastings, upon the request of Sheriff Martin, sent members of the Pennsylvania National Guard, a contingent of twenty-five hundred troops, into the area. They began arriving early in Hazleton under the command of Brigadier General J. P. S. Gobin. The community, with a population of around fifteen thousand, appeared to be under a siege with soldiers seen at every turn. The troops made daily patrols through the region until they were removed two-and-a-half weeks later.[25] Martin knew that he had lost all chance of being an effective law officer after the shooting. He also realized that sending the troops could possibly prevent legal action against him and his deputies. The mine operators welcomed the National Guard as an effective instrument in curbing strike activity, which it was.

Undaunted by the large military presence, people gathered on Saturday evening at three large indignation meetings to protest the Lattimer massacre. At Harwood, a small coal-mining village a few miles southwest of Hazleton where the doomed march began, some one thousand people gathered to register their protest in a number of resolutions regarding what had occurred the day before. They extended sympathy to the victims of the Lattimer shooting —many had been their neighbors—and spoke of being economically oppressed by the Pardee Coal Company.[26] In recounting the ill-fated march to Lattimer they declared:

> We assembled together peacefully and to seek redress for our grievances. Not one man amongst us was armed. Our mission was not to take human life nor destroy property, but to go and meet our fellow employees of the same company at Lattimer, who were in sympathy with us. We were opposed on the public highway and without provocation were shot down like dogs. That we look upon such shooting as unprovoked and uncalled for, and that if such slaughter is not murder in law it surely must be before High Heaven.[27]

A second meeting that evening in Hazle Township, just south of Hazleton, characterized the Lattimer shooting: "A great calamity which will go down in history as the greatest crime of the Christian era, has befallen this peaceful community."[28] Resolutions adopted were similar to those passed in Harwood, but made some additional demands. The people insisted that James Martin resign as sheriff and urged that the district attorney of Luzerne county prosecute those responsible for the shooting to the full extent of the law.

In Hazleton, the largest of the three meetings that evening, with some five thousand attending, was bigger than the one the previous evening. Father Richard C. Aust, priest of St. Stanislaus Polish Catholic Church, and other area clergymen, primarily representing Catholic parishes of eastern Europeans, organized the meeting. The major speaker was John Shea, a young attorney from Wilkes-Barre, who declared:

> Pulaski, a noble man of your birth, fought for our independence in 1776, he died in the service and his name is recorded in the history of your nation. Yet a lot of cowards, led by a cowardly sheriff deliberately shot into the offspring of a nation who sent one of her best sons to our clime to fight for you and I. The idea of human blood desecrated and spilled like milk by a lot of cowards and loafers.[29]

Other speakers at the Hazleton rally included various churchmen, the labor leader John Fahy, two Hazleton businessmen, John Nemeth and Matthew Long, all of whom stressed two basic ideas in their speeches: avoid illegal and disorderly acts and maintain the peace. Nemeth, a Hungarian native, implored the crowd that for its best interests it was essential that it adhere to the following: "Keep quiet, say nothing to anyone to provoke their anger, use no bad or denunciatory language; go home and stay there, go to work or stay away as you deem it wise; follow these rules and you will win your cause."[30] Aust reiterated the same idea by reminding members of the audience to remain calm and use their influence to keep the community on their side. The hope was that there would be no incidents which the coal operators and the sheriff could cite, claiming that the immigrant strikers were inclined to ruckus and riotous behavior. There was the fear that, with a sense of anger and frustration felt by so many, further violence could erupt. Any destructiveness by the strikers would undermine their cause and lessen the feelings of indignation against Sherriff Martin and his deputies. Father Stas, one of the other speakers, exhorted the audience by declaring: "We were not born in this country but we can be as good a citizen as the next one."[31]

Fahy was one of the important voices of moderation that counseled immigrant workers who felt the impulse to vent their anger for the grievous events at Lattimer. As a labor leader in the coal fields, he worked with the classes that felt exploited and manipulated. With that association, he knew their bitter feelings of frustration and, after the injuring and killing of members from their own ranks, their contemptuous attitude toward the coal operators and law officers. It would not have taken much imagination to exacerbate those passions which would have, most likely, led to violence and turmoil. Having this awareness, Fahy became one of the important proponents advocating restraint and non-violent behaviour. Speaking to a newspaper reporter he stated:

> I had advised the men to be quiet, peaceful, and not to, in any manner, violate the law, impressing upon them the necessity of all the number

being organized in one body, and then making a move along the lines of peace, fairness, and justice to all concerned.[32]

Fahy's work and conciliatory approach won high praise by a Hazleton newspaper, citing his urging of the strikers not to commit any depredation, to stop marching and present their grievances in an enlightened manner."[33] Echoing this assessment were two other newspaper accounts which reported that "he has done much to prevent further lawlessness."[34] In the weeks that followed the shooting, the ethnic community, by and large, adhered to the strategy of peace and non-violence.

The people who spoke at the Hazleton rally not only urged moderation and calm but also emphasized that the deadly deeds of the sheriff and his deputies would not go unnoticed and that justice would be done. They reminded the listeners that there were legal remedies to punish those who had committed wanton murder. Attorney Shea said, "The courts are open to you whether you can speak English or not, whether rich or poor."[36] The speeches expressed a confidence that the law officers would be held accountable for their actions, prompting many in the immigrant community to believe that the judicial system would provide a legal redress for the injustice committed at Lattimer. Before the meeting adjourned, the following resolution of condemnation was unanimously adopted:

> Resolved, That it is the sense of the thinking people of this region that the taking of human life by the Sheriff of Luzerne County and his deputies at Lattimer Friday, was wholly unwarranted and nothing less than wholesale murder, and be it further
>
> Resolved, That they should each be indicted for murder and prosecuted to the full extent of the law. And to that end be it further
>
> Resolved, That the chairman of this meeting appoint a committee to solicit funds to defray the expenses of the same. And be it further
>
> Resolved, That we, the people now here assembled, most urgently protest against the declaration of martial law as unnecessary and unwarranted, and for the sole purpose of protecting the murderous deputies from arrest by civil process.[37]

One of the important points in the above resolution was the call for creating a committee to seek the arrest and prosecution of Sheriff James Martin and his deputies. Growing out of this appeal was a meeting five days later in Matthew Long's office that led to the establishment of the National Prosecuting and Charity Committee of the Lattimer victims. This organization became the advocate and champion of the Slavic-American community to seek a legal redress of grievances against the law officers and to offer assistance to those families directly affected by the Lattimer massacre. Elected as the committee's president was Father Aust, and John Nemeth became its treasurer; of its thirteen members, four were clergymen.[38]

Aust, undoubtedly, felt a special obligation, since nine of the nineteen killed at Lattimer were members of his parish. He undertook an active role in organizing the committee to accomplish two major tasks: prosecution of Sheriff Martin and his deputies and financial support to the families of those who were killed or wounded at Lattimer. The day after the shooting, Aust sent a telegram to the Polish National Alliance Convention meeting in Philadelphia informing them of the Lattimer tragedy.[39] In response, the convention denounced what happened, describing the strikers as cattle which could be killed with impunity. It then voted to set aside one thousand dollars to help prosecute Martin and his men.[40] Within a week, Aust organized a meeting of the Slavonic clergymen in the Wilkes-Barre area to secure their aid in raising funds to support the Committee's work.[41]

One of the first acts of the Committee was to issue a formal call for support and financial contributions to help it achieve justice. The lengthy document published in the local newspaper, contained, as expected, a very critical account of the law officers' conduct. It declared that there was no riot, and labeled the sheriff and his deputies as tools of the corporations who acted "like hounds, eager for the scent . . . ready to do their master's bidding."[42] The call characterized labor conditions as worse than in Siberian Russia. It went on to state: "This country's Declaration of Independence affirms that 'all men are created free and equal.' But these deputies seemed to feel that the life of a 'foreigner' was no more precious than that of a slave."[43] Couched in patriotic terms, the appeal concluded with a request for funds.

To translate these sentiments of compassion for the victims and calls for prosecution of the perpetrators into action, the Committee needed funds. It needed money to provide assistance to the families of the injured and the dead and to retain attorneys to aid in the legal proceedings against Sheriff Martin and his deputies. Requests for contributions were repeated at the various indignation meetings; ethnic newspapers published accounts of the committee's fund-raising efforts and carried advertisements of meetings for money. *Amerikansko Slovenske Noviny,* a Slavic newspaper in Pittsburgh, ran an announcement of a special music program that would be held in Allegheny City to raise money to relieve the misery of the ill-fated survivors.[44] A month after the committee began its work, John Nemeth, its treasurer, wrote a letter to *Slovak V Amerike,* another Slavic newspaper, urging gifts of money on behalf of the widows, orphans, and the wounded victims of the shooting. Reported in the same publication was the amount of money received and how it had been disbursed.[45]

The committee, in nearly a year's time, from 11 September 1897 to 30

August 1898 collected $9,167.27.[46] (In light of inflation, this amount in today's value would be equivalent to at least $100,000.00). A gift of one hundred dollars from William Bourke Cochran, former Democratic United States representative from New York City, was used for the administrative expenses of the committee. When he sent his gift, he wrote to Aust saying: "A community which would allow such a destruction of life to go unpunished could not be considered civilized."[47] There were 534 contributions amounting to $4,355.60 for the charity fund and $4,811.67 for the prosecution fund, a division of 47.5 per cent and 52.5 per cent respectively. Contributions from outside Pennsylvania came from two foreign countries, Canada and Sweden, and twenty-two states, which accounted for $4,850.26 or 53.5 per cent of the money received. Of this amount, fifty-six per cent went to the charity fund and the remaining forty-four per cent to the prosecution fund. The five leading states sending funds were New York, Illinois, Ohio, New Jersey, and Connecticut.

Gifts raised in Pennsylvania had a different pattern: thirty-eight per cent of the money went for charity and sixty-two per cent for prosecution. Donations from Hazleton and the immediate surrounding area came to $1,023.64, or only eleven per cent of the total amount raised. Of this sum, forty-five per cent went to the charity fund and the other fifty-five per cent to the prosecution fund. A review of the list of donors from the Hazleton region to the Committee revealed that the vast majority of contributors came from the eastern European community.

Of the total amount of money collected, ninety-one per cent came from the Slavic-American community, raised by different newspapers, fraternal organizations, and parishes. The Polish National Alliance gave $500, half of what it had earlier promised,[48] and the Greek Catholic Union provided $200; these gifts went to the prosecution fund. There were few individual contributions and they were usually small amounts of less than five dollars. There were some exceptions, such as Cochran's donation, and a suprising $100 from James E. Roderick, Superintendent of the A. S. Van-Wickle collieries outside of Hazleton, specified for the charity fund.

Organized labor's monetary support was substantially less than its public pledges of support. A few days after the Lattimer shooting, Peter J. McGuire, Secretary General of the United Brotherhood of Carpenters and Joiners, declared his union would give $500 to help prosecute the sheriff.[49] The AF of L Executive Council had promised to raise funds for prosecution.[50] The committee's records, however, depict a much different picture. Neither of the two organizations made a contribution. Unions gave only $391 or 4.3 per cent of the total amount raised by the committee, and of this nearly all was designated to charity. Labor's largest donation, $200, came from the United Mine Workers of America sub-district around Hazleton. The remaining amount, $191, came from five unions in New York and Philadelphia.

The charity fund supported the basic needs of the families of the injured and dead from September 1897 until February 1899. Over this period of eighteen months, the committee cared for fifty-nine adults (widows and cripples) and thirty-eight children who lost their fathers. When the funds were exhausted, Father Aust made arrangements for the families who needed further assistance to be helped by the poor board.[51]

The other major task of the committee was to get Sheriff Martin and his deputies prosecuted. After the shooting there was a strong demand coming from the immigrant community, led by the various Hungarian societies, that they be immediately apprehended and incarcerated.[52] Brigadier General J. P. S. Gobin, Commander of the National Guard of Pennsylvania sent to occupy the Hazleton area, refused to permit their arrest. It was his position that the National Guard was to assist the Sheriff to maintain order and that it was under his orders and direction. He argued that arresting Martin and the deputies would create a situation in which there would be no one to give orders.[53] Later, Gobin admitted another reason for prohibiting the arrest. He contended that if the deputies were arrested and put in jail, "civil authorities could have been powerless to protect the lives from the fury of the mob."[54]

Finally, after nearly two weeks and after some of the tension in the area had subsided, Gobin dropped his objections. The committee hired three attorneys to work with District Attorney Fell to bring the case before a Wilkes-Barre court. This court bound the law officers over to the grand jury[55] and, a month later, returned indictments against Sheriff Martin and his deputies for the murder of nineteen and the feloniously wounding of thirty-eight strikers on September 10 at Lattimer.[56] After a couple of postponements the trial began on 1 February 1898 in the Luzerne County Courthouse, Wilkes-Barre. The District Attorney, with the assistance of the committee's three lawyers, selected as a test case the death of one of the strikers, Mike Cheslak. The trial lasted five and a half weeks and heard testimony from nearly two hundred witnesses, before the jury rendered a verdict of "not guilty" on March 9.[57]

The unsuccessful effort to win a conviction absolved Martin and his deputies of any criminal conduct in the Lattimer massacre. In *Slovak V Amerike,* a week before the verdict defeated the hopes of the Slavic community, there was an appeal that justice would not be affected by wealth and power.

> If the jury has less good sense than to see the chicanery behind all this false testimony, secured at a costly price, then the murderers will all go free, and truth will be humbled again and the whole world will see that even in this free country of America the poor cannot be assured of the shield of truth.[58]

An editorial in *La Questione Sociale,* an Italian-American newspaper, denounced the acquittal. It said that it was difficult to "believe that the American middle class could reach the point to overlook and approve an infamy. It reached the point where the word justice is meaningless, brutality and cruelty are tolerable."[59]

At the outset of the trial, there were expressions of doubt in the Slavic community that Sheriff Martin and his deputies would be convicted. In *Slovak V Amerike,* an editorial suggested that the sheriff and deputies would get a "velvet-gloved" acquittal or, if found guilty, they would most likely receive an easy sentence. The reason for this attitude, according to the writer, stemmed from a feeling that there was "much public sentiment against the 'Hungarian' element, but most significantly, a verdict charging Martin as guilty would mean that Luzerne County stands to pay court's fees, reparation, and expenses."[60] It was readily acknowledged that nativist sentiments against the Slavic community were widely held in the anthracite region. At the time of the trial it was reported: "No man can be found in Wilkes-Barre who will speak a good word for them."[61] The fear of bigotry prompted this concern about the jurors: "Are their hearts filled with rank nativism and malicious ill-will against foreigners? If they are bigoted nativists, it is vain to expect a fair sentence."[62] The question seemed pertinent to the ethnic community when the members of the jury were selected: men of English heritage, none employed in mining, none from the Hazleton area.[63] With the hopes of the Slavic-American community for a judicial redress of the wrongs committed at Lattimer shattered with the news of the acquittal, *Amerikanski-Slovenske Noviny* printed an editorial that began: "They will kill, they will shoot, and grant you no mourning or tears of relief. And somehow they will make it all come out within the limits of the law."[64] In the committee's final report, Aust wrote:

> The fight was made and unluckily lost, the case and the whole proceeding in the case will go down to future, in the unwritten history of suffering manhood, as another instance where, at least in the eyes of the laymen, if not professional lawyers, capital and influence, and not true justice won the day. May God be the father and protector of the poor unfortunate ones; may He be the avenger of the unfortunate victims.[65]

In Hazleton, as well as in other parts of the country, there was a deep concern over the tragedy that occurred at Lattimer, especially within the immigrant population. It is important to note that violence, which took so many lives and injured even more, was not followed by more violence. Part of the answer for this lies in the important role that a number of clergymen, labor leaders, and others took as a moderating influence against any illegal or provocative actions. The immigrant community basically adhered to the advice of their leaders. This is not simply to dismiss the influence which the

National Guard had by patrolling the area, but their role was made much less difficult by the attitudes of the leaders who called for moderation and nonviolence, and instilled a belief that justice would prevail over what was seen as a gross act of injustice. That faith was nurtured by the National Prosecuting and Charity Committee of the Lattimer Victims. It proved to be a valuable organization in seeking a judicial remedy to the resentment over the actions of Sheriff Martin and his deputies, provided support to those who were made destitute, and served as an advocate for the powerless. For the eastern European immigrant community that suffered the sorrow of Lattimer, the committee was a well-spring of hope.

The committee was not the only advocate of the Lattimer victims; the Austro-Hungarian government also assumed this role. When the Austro-Hungarian consulate at Philadelphia learned of the incident, it dispatched its secretary, Dr. Thodorovich, to the scene on September 11 to undertake an investigation.[66] His main purpose was to secure first-hand information on what occurred at Lattimer so that his government would be well-informed without having to rely upon newspaper accounts. Arriving in Hazleton, he contacted leaders of the Slavic community to aid him in his inquiry; John Nemeth assisted in locating and interviewing eye-witnesses to the shooting. After spending a week in the area, he specifically identified ten of the nineteen killed and eleven of the thirty-eight wounded as Austrian and Hungarian citizens.[67] Later, an American government official concluded that in most instances the Lattimer casualties were primarily of Hungarian or Polish origin and were Austro-Hungarian subjects.[68]

Speculation began to grow that the Austro-Hungarian government would lodge a diplomatic protest with the United States government and seek an indemnity for the law officers' attack on its citizens.[69] This arose in part from similar occurrences during the past ten years involving nationals of China and Italy. Chinese immigrants in Rock Springs, Wyoming, were subjected to nativist mob violence in 1885 with twenty-eight murdered, followed by other riots that destroyed the property of Chinese citizens in Seattle and Tacoma, Washington. In response to these disgraceful attacks, that went unpunished, and a desire to secure better diplomatic relations with China, the Congress, in 1887, paid the Chinese government an indemnity of $424,368.49.[70] The second incident involved Italians in New Orleans in 1891. The city's chief of police had been murdered, with the public strongly suspecting that members of the Italian community were responsible. None of those tried was found guilty; whereupon, a roaring mob broke into the jail and removed eleven people of Italian origin and lynched them. President Harrison, in his state of the union message, offered an apology to the Italian government and directed Secretary of State Blaine to make a payment of $25,000.00 from the department's contingency fund to the families of the victims.[71]

Pressure from the American Slavic community on the Austro-Hungarian government to do something began to emerge. The Polish National Alliance, meeting in convention at Philadelphia, passed a resolution saying in part that at Lattimer

> a large number of miners in the Hazleton district have been attacked, killed and wounded by armed deputy Sheriffs, . . . who at the time of the attack were peacefully marching, and were not engaged in any unlawful or riotous pursuit, . . . we condemn the hasty action of Sheriff James Martin, of Luzerne County, and his deputies, and their murderous attack perpetrated upon a public highway without justification or excuse.[72]

John Spivak, Grand Secretary, First Catholic Slovak Union of America, declared: "The slaughter near Hazleton was such a cruel and cowardly act that it fills every honest and decent citizen of this country with horror."[73] Austro-Hungarian nationals in Chicago expressed their indignation two days after the Lattimer massacre.[74] A mass meeting of Poles and Slovaks held in New York passed a series of resolutions saying in part: "As American citizens, as workingmen, as Slavs by accident of birth, [we] do most emphatically denounce and condemn this criminal act."[75] There were expressions of condolence and pledges of material assistance to the bereaved families. Professing their faith in the American justice system, those at the mass meeting demanded that those who were blameworthy be tried. In addition, there should be monetary compensation in an effort to make restitution to the survivors of the victims' families.

There were calls from around the country for the Austro-Hungarian government to seek a redress for the injustice done at Lattimer. A group of Austrian-Polish citizens in Chicago sent a message to the Premier, Count Badent, and Minister of Foreign Affairs, Goluchowski, appealing to the Austro-Hungarian government "to institute proceedings to secure reparations to bereaved families and punishment of perpetrators of outrage."[76] In New York, Emil Nyitray, President of the Hungarian Reformed Church, advocated the following: "The representative of the Austro-Hungarian government should certainly see to it that the families of those killed and disabled get damages."[77]

An editorial in *Amerikansko-Slovenski Noviny* made critical comments about the ineffectiveness of the Austro-Hungarian consuls who were duty bound to represent Slavic interests in this country. In contrast, it called attention to the success of the Italian diplomats who defended their countrymen in the New Orleans affair. "It is our reasoned judgment that the Hazleton bloodshed is far more criminal than was the New Orleans lynching and it is the patent duty of the Austro-Hungarian government to make honorable and serious intervention on our behalf."[78]

In a letter from the Philadelphia Consul, Alfred J. Ostheimer, to Thodorovich that appeared in the Wilkes-Barre newspaper, he urged him to "collect all the evidence we can possibly get to make our case as strong as possible, and I sincerely trust that you will succeed in getting it."[79] In the same correspondence, Ostheimer expressed his own conviction that the sheriff and posse were criminally liable for their unwarranted actions. He also referred to the United States paying claims to foreign countries, as much as fifty thousand dollars per individual when their nationals had lacked protection and been attacked and killed. He urged Thodorovich to be diligent in collecting affidavits and as much evidence as possible so that the Austro-Hungarian government could successfully press its case.

With this directive, Thodorovich secured twelve affidavits, nine from strikers who marched from Harwood to Lattimer, and three from individuals at Lattimer who observed the confrontation between the law officers and the strikers. He also obtained a transcript of a mortician's testimony at the corner's inquest which described the mortal wounds of eleven victims: nine had been shot in the back. The depositions of the strikers and observers declared that the marchers were unarmed, were peacefully using the public highway, and were halted by the sheriff, whereupon the deputies, without provocation, began shooting for at least two minutes resulting in a carnage of death and injuries.[80]

As Thodorovich was concluding his investigation, a newspaper quoted his personal opinion of the shooting, "I have no opinion to express concerning Sheriff Martin in an official way, but personally, of course, I have formed an idea which does not excuse him as an official or an individual."[81] This comment, Consul Ostheimer's request for a well-documented report, and the outpouring of indignation, particularily among eastern European immigrants, prompted an area newspaper editorial to comment on the possibility that the Austro-Hungarian government might seek an indemnity from the American government.

> It would even be in order to serve notice on those Austrian officials that they are making themselves unnecessarily conspicuous as well as impertinent when they assume to designate the killing of the Hungarians at Lattimer as "bloody butcher" and "wanton murder." It remains yet for the coroner to hold an inquest, and for a court and jury to try the accused officers before anyone can justly charge that murder was committed.[82]

A more critical editorial, reflecting a nativist outlook charged that the marchers attacked the law officers and were in a state of violent revolt against the laws of the State and the lawful authority of the county. It further asserted that it would be erroneous to believe that the event raised any international question. "If any subjects of the King of Hungary believe that they are not fairly treated in Pennsylvania, they are at liberty to withdraw. Pennsylvania could bear with composure the deportation of the whole Hungarian population of Luzerne county."[83]

Within two weeks, Thodorovich submitted his report, which became the basis of a formal diplomatic note from the Austro-Hungarian government to the United States expressing deep concern over the killing and wounding of its nationals at Lattimer. In registering his government's distress, Minister Hengelmuller stated that his country

> can not avoid the impression that its subjects suffered death or wounds, not in consequence of unlawful resistance to the constituted authorities, and therefore not through their fault or owing to an unfortunate accident, but through an unjustifiable, illegal, and, as it appears, improper use of official authority of the sheriff, consequently of a responsible representative of the authority of the State.[84]

He requested that the federal government make an investigation of the facts surrounding the Lattimer shooting and inform him as soon as possible of its findings. The message concluded by declaring that the Austro-Hungarian government reserved the right to seek an appropriate indemnity for its subjects who were killed or injured at Lattimer and for their surviving relatives. Two weeks later, Secretary of State John Sherman sent a short note pledging to give the matter "prompt and careful attention demanded by the gravity of the matters set forth in your communication."[85] These initial exchanges began a long and protracted debate between the two governments over the Lattimer massacre, lasting nearly two years.

Sherman, seeking to honor his commitment to Hengelmuller, wrote to the Pennsylvania Governor, Daniel Hastings, asking him to supply the State Department with all the pertinent information concerning what happened at Lattimer. Hastings assured Sherman that he would promptly report the relevant facts to the department. What transpired instead was a series of delays by Hastings in sending the requested material. As Hengelmuller continued to press the State Department in November and December for its report, Sherman in turn urgently repeated his appeal to Hastings to forward the information. At first Hastings explained his tardiness by saying that he did not yet have the reports from General Gobin of the Pennsylvania National Guard or from the Luzerne County Sheriff, James Martin.[86] Finally, after three months, Hastings revealed the basic reason for the delay when he wrote Sherman. Since the Sheriff and his deputies had been indicted on October 27 for murder and were shortly to be tried, their counsel wanted assurances that the information in the reports from Gobin and Martin would not be made public before the trial, since it might be prejudicial to their case.[87]

Sheriff Martin's lengthy letter to Governor Hastings written on October 18 chronicles the events leading up to the shooting and what happened when the confrontation occurred at Lattimer. There was one passage that could possibly cause some detrimental reactions if made public prior to the trial. Martin clearly indicated he was under pressure from the Pardee Coal Company to prevent the Lattimer mines from being shut down.

> The reason I stopped the mob before they came to the breaker and mines, was that I was notified by Supt. Mr. Drake that they would hold me responsible for any interference with their mines. I knew very well that if the mob got to the breaker, they would surely attempt to stop it, and then if we undertook to arrest them that there would be serious trouble, and I have no doubt that some innocent blood would have been shed, and there is no doubt in my mind that the loss of life would have been greater.[88]

In concluding his letter he objected to the Austro-Hungarian government's intervention into the matter.

> I do not think it right and fair for any foreign country to be allowed to try and prejudice our case at the present time, and furthermore, if our act and actions were wrong, our courts and the American people are the proper parties to say so, and they will not be afraid to do so, without the interference of the Austrian government.[89]

Martin's acknowledgment that he was under pressure by the Pardee Coal Company to prevent a shutdown was reinforced in Gobin's report to Hastings in which he indicated that the coal company had applied to the sheriff for protection. He stated that the sheriff approached the strikers at Lattimer with a drawn revolver. The report raised the spectre that some of the deputies fired their weapons more than once. It did not indicate that the deputies fired any warning shots nor was there any statement that the sheriff ordered a halt to the shooting. Gobin characterized the foreign population in the Hazleton area as a violent people who carried weapons and frequently fought and killed each other.[90]

Sherman asked Hastings to forward the reports from Martin and Gobin, and he assured him that they would not be made public before the trial. He then informed Hengelmuller that the promised report of all the facts surrounding the Lattimer shooting would not be available until after the trial of the law officers.[91] Hengelmuller, by then exasperated, made a lengthy and critical response to Sherman over the refusal of the American government to provide the promised report. He complained that the United States failed to keep its word and intimated that the government was impotent in carrying out any investigation. The fact that the sheriff and his deputies were to be tried was immaterial; the basic issue raised in the Lattimer shooting was the safety of Austro-Hungarian nationals in the country.

> I do not know, moreover, what is the technical basis of the indictment of Sheriff Martin and his posse, but my Government can in no case consider the technical question of his guilt or innocence of the crime with which he is charged as being synonymous with the question whether those victims of the catastrophe, who had a right to their protection, are entitled to indemnity or not.[92]

Dismissing the forthcoming trial as irrelevant, he told Sherman that his government had instructed him

> to declare that it must hold the Federal Government responsible for the injury suffered by its subjects on the occasion of the occurences in question, and that it must ask of it a just and adequate indemnity for the victims of the catastrophe who were mentioned in my note of September 28.[93]

No amount was mentioned in the request for an indemnity, nor was there any implication that diplomatic relations would be reduced or broken. What was asked was that the question of an indemnity be accepted by the United States with the amount open for negotiations.

The decision to request an indemnity resulted from a number of factors. Continued delay by the State Department to provide the report of a promised inquiry, after frequent requests, had provoked the Austro-Hungarian government to take a more affirmative position to show its displeasure. There was the expectation of the Slavic-American community that the Austro-Hungarian government would be its advocate, particularly in light of nativist sentiment and because it was the victim of an attack. In the Hapsburg empire, it was important to demonstrate to its citizens its concern and a resolve to seek redress from a foreign country in which its nationals, it was believed, had been killed and wounded, without justification, by over-zealous law officers. National esteem required it. Had not the Americans in the last few years paid out awards to the Chinese and Italians who lacked protection and had been attacked and killed? The Austro-Hungarian government felt it important to seek damages prior to the trial so that the Americans could not use the verdict to decide if it would pay.

The American State Department rejected the indemnity request outright on the grounds that it was still an undetermined fact whether the law officials acted within their lawful limits or acted improperly. Sherman declared, "The essential question is whether the degree of force employed was or was not lawful."[94] The answer to this question would not be decided by the Austro-Hungarian government, but rather in a Pennsylvania court of law.

The position was further clarified after Sherman studied the reports sent from Governor Hastings. Writing to Hengelmuller, Sherman gave his opinion of what transpired at Lattimer and expressed some doubt that the strikers were peacefully assembled. "The facts would rather appear to be that, upon the sheriff advancing unattended in order to meet the ringleaders, he was dangerously assaulted and that shots were fired, without command, by the deputies in the attempt to rescue him."[95] However, the Secretary of State promised to suspend final judgment until the trial had reached a verdict.

It was apparent that the State Department had come to view the trial of Sheriff Martin and his deputies as the primary factor in the indemnity issue. With the trial so crucial, the State Department decided that it would be prudent to send a representative of the Justice Department to Wilkes-Barre to attend the trial and report on its proceedings. With a courtroom observer the government's ability to develop a sound position on the indemnity question would be enhanced. In addition, it would give a positive response to the Austro-Hungarian government that the Americans were sincerely interested in ascertaining what actually occurred at Lattimer. After Hengelmuller learned of this decision, he expressed warm appreciation and satisfaction; in addition, he indicated that he would forego discussing the merits of the case until the trial was over.[96]

Appointed as the State Department's observer was Assistant Attorney General, Henry Hoyt, the son of former Pennsylvania Governor, Henry Martin Hoyt, and a native of Wilkes-Barre. The Austro-Hungarian government, deciding that it was essential to have its own representative at the trial, sent Attorney Robert D. Coxe of Philadelphia to report on its proceedings. The fact that both countries sent representatives to the Luzerne County court dramatized the importance placed upon the outcome of the trial. When the case began, a local newspaper declared:

> The momentous trial is on, the trial which is to decide so many interesting questions, national and international,—which is to decide whether or not a body of strikers may have the right of way on a public highway,—which is to decide whether or not indemnity shall be paid to the foreign governments for the killing of certain of their citizens resident in this country, —which is to decide whether or not the sheriff of Luzerne County and his posse are guilty of murder and felonious wounding.[97]

The central legal issue in the case was whether or not there existed in the Hazleton area, due to the actions of the strikers, a riotous condition. The prosecution argued that the public order and safety were not threatened by the strikers, who were peacefully using a public highway when they were shot by the deputies at Lattimer. No public disorder existed nor threat of a riot in the Hazleton area to warrent the legal establishment of a *posse comitatus,* and, therefore, its existence was illegal. The prosecution further maintained that this illegitimate body then acted in an unjustifiable and unnecessary manner and, consequently, committed a criminal act of murder. The defense argued that the conditions justified calling out the *posse comitatus.* This was its key point. The trial judge, Stanley Woodward, instructing the jury, drew its attention to the controverted testimony on this issue and declared that it would have to decide if riotous conditions existed. He also stated that if

> it was the right of the sheriff to command the crowd to disperse, then it was the duty of the crowd to obey his command. The right to give the order implies the duty of obedience to the order, and disobedience of it is evidence of a riotous purpose. If I push on when the sheriff orders me to stop I do so at my own peril.[98]

The jury, upon evaluating the evidence, believed that a riotous condition did in fact exist. With that issue settled, the jury concluded that the actions of the deputies which resulted in the killing and wounding of some of the strikers were free from malicious intent and were legally required to maintain law and order.

Henry Hoyt, without reservation, in his report of the trial, indicated:

> There is no question in my mind that the court ruled fairly as to the admission of evidence and upon the various points arising throughout the trial, nor can it be denied, I think, that the charge of the court was full, fair, and sound, and stated the law as settled by the course of Anglo-Saxon jurisprudence for several hundred years, under statutory as well as under common law, correctly and without failing to do entire justice to the respective contentions of the prosecution and the defense.[99]

Feeling that the case was fairly tried and a just verdict delivered, he rejected the idea that those who were wounded and the families of those killed should be entitled to any indemnity. It was Hoyt's contention that the strike leaders purposely sought to provoke disorder and to encourage hostility to the more prosperous classes and the interests of capital.[100] Having this attitude led him to conclude that the conflict at Lattimer was unavoidable as long as the sheriff and his posse were determined to preserve civil order and respect to the law. Hoyt said "that under all the circumstances the action of the sheriff and the posse, although fatal and lamentable in its result, was clearly justifiable."[101]

Not to completely shut the door for some kind of remedy, Hoyt suggested that the foreigners could institute a civil suit in either a state or federal court for damages. If they were successful in establishing the liability of the sheriff and his deputies, then, and only then, could an indemnity from the American government be considered.[102]

The State Department accepted Hoyt's assessment of the trial and his recommendations regarding the indemnity issue. In mid-April, Secretary of State William Day sent Hoyt's report to Hengelmuller and declared that there was no basis for the American government to honor any claims of indemnity. As expected, the Austro-Hungarian government completely rejected the report, maintaining that the question of indemnity was an issue to be decided by negotiations and not by a trial in the Luzerne County court.[103] To underscore his government's position, he cited a report prepared by the legation's counsel, Robert D. Coxe, who witnessed the trial.

The Coxe report was unequivocal in asserting that "the trial resulted in a miscarriage of justice"[104] and "in the absence of any defense, it would unquestionably have been the duty of the jury to have convicted the defendants, . . . to have supported a claim for compensation on behalf of the victims of the Lattimer shootings September 10, 1897, and their representatives."[105] In his assessment of the trial Coxe argued that the jury was not impartial nor truly representative of the community. He disputed the contention that the Hazleton area was in a state of public disorder or the threat of riotous conditions existed and claimed the major reason for organizing a *posse comitatus* was simply to serve the interests of the coal operators in order to break the strike. He was convinced that the deputies were victims of panic and had no justification for shooting the strikers.

Disregarding the jury's verdict and maintaining that a grave injustice occurred at Lattimer, Coxe asserted that there was a past precedent in which the American government gave monetary restitution to foreigners who were victims of American violence solely on the basis of generosity and pity. In the Chinese indemnity case he said, the Congress had based its decision on a benevolent spirit rather than a legal obligation.[106]

In the summer of 1898, the Austro-Hungarian government continued its efforts to get the State Department to alter its stand on the indemnity issue. In Vienna, the foreign ministry called the American minister, Charlemagne Tower, to its offices to express its disappointment and disagreement with the United States position. It felt that the court's decision should not preclude the two countries from entering into diplomatic negotiations to resolve the dispute. It requested that the Americans be guided by the principles of charity and humanity and reverse its refusal.[107] In Washington, Hengelmuller repeated the request but to no avail. The case remained in limbo for the rest of the year.

During this time, Baron Riedl replaced Hengelmuller as the Austro-Hungarian Minister and John Hay became the new Secretary of State. Hay, in January 1899, agreed to an interview with Attorney Coxe to discuss the case. After their meeting, Hay wrote to Baron Riedl and set forth in explicit language the American government's definitive position on the indemnity issue with the hope of finally settling it. Reiterating many of the points made by his predecessor, Secretary Sherman, Hay emphasized that a Pennsylvania court had found Sheriff Martin and his deputies' actions within the limits legally permitted in discharging their duties as the conservators of peace. The strikers were guilty of disturbing the public peace and violating the law. The trial was fair and there was no denial of justice.[108] Clearly expressing what the United States government's position was he declared:

> This Government recognizes the international obligation to do justice, but it can not admit that in this case legal injustice has been done. Even if it were to be conceded that the sheriff and his deputies were acting wrongfully and unlawfully, still the remedy by way of diplomatic intervention can not be invoked until all remedies have been exhausted before the ordinary judicial tribunals. In this case abundant remedies are afforded for redress, if any actionable wrong has been committed; but the disposition of this claim may safely be rested on higher grounds—on the ground that aliens are subject to the same rules of law and order, of peace and justice which bind the citizens of the United States. Whoever sojourns in a foreign land having a settled and pure administration of justice impliedly submits to the local jurisdiction and to the requirements of the municipal law. This Government can not tolerate a state of anarchy, either of its own citizens or of aliens who may engage in industrial or other pursuits within its territory. If they obey the precepts of the law it will protect them; if they defy the law and the constituted authorities, then, in common with all others who participate with them in such acts of lawlessness and violence, they must be deemed to accept the consequences of the conflict which they invite.[109]

Without doubt, the United States refused to admit that it had any liability or obligation to pay an indemnity for what happened at Lattimer.

Nearly three months lapsed before Riedl's replacement, Baron Von Riedenau, responded to Hay's message of rejection. The Austro-Hungarian government refused to agree with the American position and continued to argue that the trial was irrelevant to the legitimate issues raised in its demand for an indemnity. Irrespective of the trial, the Austro-Hungarian government believed its citizens were killed and wounded by unlawful actions of the sheriff and his deputies. In an effort to find a possible way to resolve the dispute Von Riedenau proposed that the indemnity issue be submitted to a court of arbitration for settlement. He reinforced his proposal with two arguments. First, their differences centered only on conflicting legal opinions and by using a court of arbitration both governments could reach a settlement without sacrificing their dignity. Secondly, he reminded the United States of her historic position as an initiator and advocate of arbitration in resolving disputes. Accepting this approach would serve as a good example for encouraging European nations to support arbitration.[110] In advancing this argument, the Austro-Hungarian government placed the United States in an awkward situation if it refused; it hoped to capitalize on American public sentiment that supported arbitration. During this time an American delegation was at the First Hague Peace Conference giving its support to the use of arbitration to resolve international disputes.[111]

Despite this diplomatic maneuver, Hay reaffirmed the United States' opposition to any indemnity and rejected the arbitration offer. He declared, "While the Government of the United States has been a conspicuous advocate of the principle of arbitration where properly applicable, it is not believed that it applies in a case which, on the facts and on principles of public law, seems to this Government to be without foundation in justice."[112] With his reply, after nearly twenty-two months, the question of an indemnity arising out of the Lattimer shooting came to an end in June 1899.

The diplomatic incident between the two countries did not become a major confrontation. The Austro-Hungarian government avoided bellicose language in setting forth its position to Washington. There were no threats to recall its diplomats or to impose trade sanctions. The Hapsburg empire was a minor power in Europe while the United States was a rising world power, particularly with its recent victory in the Spanish-American War. Still the Austro-Hungarian government felt an obligation to be an advocate for its citizens in a foreign nation. Failure to do so would have caused domestic political problems. Besides a nationalistic obligation, there was a sincere belief that Sheriff Martin and his deputies at Lattimer misused their authority, causing unwarrented deaths and injuries. It further argued that a fair trial did not occur. In the final analysis, persuasion was the only choice the Austro-Hungarian government had in trying to convince the United States to agree to an indemnity. *The New York Times* cited the view expressed in a Vienna newspaper, *Allegemeine Montage Zeitung,*

> that the legal standpoint adopted as the reason for refusal deprives this of any offensive or hostile character . . . it is inconceivable that the reply of the United States Government should lead to any retaliatory measures on the part of Austria or to conflict between the two powers.[113]

Once the United States took the position that the trial decision would be the crucial factor in deciding whether or not to award an indemnity, negotiations with the Austro-Hungarian government on the questions became impossible. After the jury exonerated Sheriff Martin and his deputies of wrongdoing, it became politically impossible for the State department to consider an indemnity or to agree to arbitrate the matter. To have agreed to a negotiated settlement granting monetary damages would have repudiated the trial and acknowledged that the law officers' actions were less than legal. In the Chinese and Italian cases, those who were responsible for the deaths, injuries, and destruction of property were never brought to trial. To the United States there were fundamental differences between those incidents and what transpired at Lattimer and afterward. Both governments believed their responses to the Lattimer massacre were correct, even though they were diametrically opposed. Each side reflected the views of the protagonists at Lattimer. Despite the role of advocate that was taken by the National Prosecuting and Charity Committee of the Lattimer Victims and the Austro-Hungarian government in seeking a redress of what the Slavic-American community felt was a grave injustice at Lattimer, their hopes gradually diminished and they were eventually left with nothing but the memory.

NOTES

[1] Henry Hoyt, Assistant Attorney-General, to William Day, Secretary of State, 8 April 1898, *Papers Relating to the Foreign Relations of the United States, with the Annual Message of the President Transmitted to Congress December 6, 1898* (Washington, 1901), 82–87. A great deal of confusion and inaccuracy existed in newspaper accounts as to the number of casualties from the shooting; reports varied from fourteen to as many as fifty people killed. Hoyt based his figures upon grand jury indictments against Sheriff Martin and his deputies who were brought to trial on 1 February 1898.

[2] *The Daily Standard* (Hazleton), 11 September 1897.
[3] *New York Journal and Advertiser*, 12 September 1897.
[4] *Straz* (Scranton), 18 September 1897.
[5] *Svoboda* (Mt. Carmel, Pennsylvania), 16 September 1897.
[6] *Straz* (Scranton), 25 September 1897.
[7] *New York Journal and Advertiser*, 13 September 1897.
[8] *Ibid.*, 14 September 1897. Reverend Pribyl had recently moved from Hazleton to Bridgeport, Connecticut. St. Joseph's, organized in 1882, is the oldest Slovak Roman Catholic church in the western hemisphere.
[9] *The New York World*, 13 September 1897.
[10] *The Daily Standard* (Hazleton), 11 September 1897.
[11] *Pottsville Republican*, 11 September 1897.
[12] *Wilkes-Barre Record*, 11 September 1897.
[13] *Daily Globe* (South Bethlehem), 11 September 1897.
[14] *The New York Tribune*, 11 September 1897.
[15] *Detroit Free Press*, 11 September 1897.
[16] *Boston Daily Globe*, 11 September 1897.
[17] *St. Louis Post Dispatch*, 11 September 1897.

[18]*Pittsburgh Dispatch,* 11 September 1897.
[19]*San Francisco Chronicle,* 11 September 1897.
[20]*Amerikansko Slovenski Noviny* (Pittsburgh), 16 September 1897.
[21]*Ibid.*
[22]Konstantin Culen, "The Lattimer Massacre," *Slovakia,* XXVII (1977), 45–46; Konstantin Culen, *History of the Slovaks in America* (Bratislava, 1942), Vol. 1, Part XI, 162–176. Translated by Sister M. Martina Tybor, SS.C.M.
[23]*Chicago Tribune,* 13 September 1897; *The New York World,* 14 September 1897; *The New York Tribune,* 17 September 1897; *The New York Times,* 19 September 1897; *Scranton Times,* 13, 15 & 20 September 1897; *Pottsville Republican,* 13 September 1897; and *Wilkes-Barre Record,* 17, 18, 20, 23, 25, & 28 September 1897.
[24]*The Daily Standard* (Hazleton), 11 September 1897.
[25]"Report of Adjutant General," 11 January 1898, in *Commonwealth of Pennsylvania,* 1897, *Official Documents, Comprising the Departments and Other Reports, Made to the Governor, Senate and House of Representatives of Pennsylvania* (Harrisburg, 1899), Vol. III, 121–126.
[26]*Scranton Times,* 13 September 1897.
[27]*Ibid.*
[28]*The Daily Standard* (Hazleton), 13 September 1897.
[29]*Ibid.*
[30]*Ibid.*
31*Ibid.*
[32]*Pottsville Republican,* 13 September 1897.
[33]*The Daily Standard* (Hazleton), 17 September 1897.
[34]*Daily Miners' Journal* (Pottsville), 14 September 1897.
[35]*Hazleton Piain Speaker,* 23 September 1897.
[36]*The Daily Standard* (Hazleton), 13 September 1897. Even within the English-speaking population, Rev. J. B. Waines of the Hazleton English Lutheran Church emphasized a reliance on the court system in a sermon to his congregation. "Rigid investigation there will be by the civil authorities created and sanctioned by law. In due time and by due process the awful responsibility will be fixed." The minister further asserted that a "righteous law will award a righteous penalty." (*New York Journal and Advertiser,* 13 September 1897.)
[37]*Ibid.*
[38]*Ibid,* 17 September 1897.
[39]*Philadelphia Inquirer,* 16 September 1897.
[40]*Zgoda* (Chicago), 16 September 1897.
[41]*Pottsville Republican,* 18 September 1897. When Father Aust died in 1913, the local newspaper wrote: "It is no idle boast when we say that of all the priests of the foreign congregation none was more beloved than he." (*The Daily Standard* [Hazleton], 24 September 1913.) At the same time John Mitchell, former President of the United Mine Workers of America, sent a telegram stating: "In the death of Rev. Father Aust, the miners of Pennsylvania have lost a valuable friend and the community of Hazleton a good citizen." (*The Daily Standard* [Hazleton], 27 September 1913.)
[42]*Ibid.,* 24 September 1897.
[43]*Ibid.*
[44]*Amerikansko Slovenske Noviny* (Pittsburgh), 16 November 1897.
[45]*Slovak V Amerike* (New York), 14 October 1897.
[46]"Report of the National Prosecuting and Charity Committee of the Lattimer Victims" (Freeland, 1899).

⁴⁷*Scranton Times,* 7 March 1898.
⁴⁸*Zgoda* (Chicago), 16 September 1897.
⁴⁹*Wilkes-Barre Record,* 14 September 1897. Many of the deputies fearful of reprisals, temporarily left the Hazleton area following the shooting. Those who remained often had soldiers guarding their homes. It was not unusual for an individual to put a notice in the newspaper disclaiming that he was a member of Sheriff Martin's posse or that he had been at Lattimer. Several days after the tragedy the hostility of the working class against the deputies was still intense. (*The Daily Standard* [Hazleton], 14 & 16, September 1897; *Pottsville Republican,* 14 September 1897; *Scranton Times,* 15 & 16 September 1897; and *Philadelphia Inquirer,* 16 & 19, September 1897.)
⁵⁰*Washington Post,* 24 September 1897.
⁵¹"Report of the National Prosecuting and Charity Committee of the Lattimer Victims," 1.
⁵²*Wilkes-Barre Leader,* 12 September 1897.
⁵³*Philadelphia Inquirer,* 14 September 1897.
⁵⁴*Pottsville Republican,* 21 September 1897.
⁵⁵*Wilkes-Barre Record,* 23 September 1897.
⁵⁶*Ibid.,* 28 October 1897 and George B. Kupl, ed., *Luzerne Legal Register Reports* (Wilkes-Barre, 1900), Vol. IX, 70.
⁵⁷*The Record of the Times* (Wilkes-Barre), 11 March 1898.
⁵⁸*Slovak V Amerike* (New York), 3 March 1898.
⁵⁹*La Questione Sociale* (Paterson, New Jersey), 15 March 1898.
⁶⁰*Slovak V Amerike* (New York), 3, February 1898.
⁶¹*The New York World,* 13 February 1898.
⁶²Culen, "The Lattimer Massacre," 53.
⁶³*The Record of the Times* (Wilkes-Barre), 4 February, 1898.
⁶⁴Culen, "The Lattimer Massacre," 57.
⁶⁵"Report of the National Prosecuting and Charity Committee of the Lattimer Victims," 1.
⁶⁶*Wilkes-Barre Leader,* 11 September 1897.
⁶⁷*Philadelphia Inquirer,* 13 September 1897. Hengelmuller, Austro-Hungarian Minister to the United States, to John Sherman, Secretary of State, 28 September 1897, in *Papers Relating to the Foreign Relations of the United States,* 46.
⁶⁸Hoyt to Day, 19 April 1898, *Ibid.,* 101.
⁶⁹*The New York World,* 13 September 1897; *Wilkes-Barre Record,* 15 September 1897; *Philadelphia Evening Bulletin,* 13 September 1897; *The Record of the Times* (Wilkes-Barre), 17 September 1897; and *L'Italia* (Chicago), 18 September 1897.
⁷⁰Charles C. Tansill, *The Foreign Policy of Thomas F. Bayard* (New York, 1940), 138, 149, and 159.
⁷¹David S. Muzzey, *James G. Blaine: A Political Idol of Other Days* (Port Washington, New York, 1934), 411–413.
⁷²*Philadelphia Inquirer,* 12 September 1897 and *Zgoda* (Chicago), 16 September 1897.
⁷³*The New York World,* 14 September 1897.
⁷⁴*The New York Times,* 12 September 1897.
⁷⁵Culen, "The Lattimer Massacre," 50.
⁷⁶*The New York Times,* 12 September 1897.
⁷⁷*The New York World,* 14 September 1897.
⁷⁸Culen, "The Lattimer Massacre," 48.
⁷⁹*The Wilkes-Barre Record,* 16 September 1897.

⁸⁰Hengelmuller to Sherman, 28 September 1897, *Papers Relating to the Foreign Relations of the United States,* 47–56.

⁸¹*The Record of the Times* (Wilkes-Barre), September 17, 1897.

⁸²*Wilkes-Barre Record,* 17 September 1897.

⁸³*The Philadelphia Times,* 15 September 1897.

⁸⁴Hengelmuller to Sherman, 28 September 1897, *Papers Relating to the Foreign Relations of the United States,* 47.

⁸⁵Sherman to Hengelmuller, 9 October 1897, *Papers Relating to the Foreign Relations of the United States,* 56

⁸⁶Sherman to Hastings, Governor of Pennsylvania, 11 October 1897; Sherman to Hengelmuller, 12 October 1897; Hastings to Sherman, 13 October 1897; Hengelmuller to Sherman, 11 November 1897; Sherman to Hastings, 12 November 1897; Sherman to Hengelmuller, 12 November 1897; Hastings to Sherman, 17 November 1897; Sherman to Hastings 10 December 1897; Sherman to Hengelmuller, 13 December, 1897; Sherman to Hastings, 22 December 1897; and Hastings to Sherman, 22 December 1897, *Papers Relating to the Foreign Relations of the United States,* 55–60.

⁸⁷Hastings to Sherman, 24 December 1897, *Papers Relating to the Foreign Relations of the United States,* 61.

⁸⁸James Martin, Sheriff of Luzerne County, to Daniel Hastings, Governor of Pennsylvania, 18 October 1897, Governor of Pennsylvania, 18 October 1897, Governor Hastings Papers, MG-145, Box 18, Division of Archives and Manuscripts Pennsylvania Historical and Musem Commission, Harrisburg, Pennsylvania. (Hereafter cited as PHMC).

⁸⁹*Ibid.*

⁹⁰J. P. S. Gobin, Brigadier General of the Pennsylvania National Guard, to Daniel Hastings, Governor of Pennsylvania, 20 October 1897, Governor Hastings Papers, MG-145, Box 18, PHMC.

⁹¹Sherman to Hastings, 28 December 1897, and Sherman to Hengelmuller, 28 December 1897, *Papers Relating to the Foreign Relations of the United States,* 61–62.

⁹²Hengelmuller to Sherman, 30 December 1897, *Papers Relating to the Foreign Relations of the United States,* 64.

⁹³*Ibid.*

⁹⁴Sherman to Hengelmuller, 8 January 1898, *Papers Relating to the Foreign Relations of the United States,* 66.

⁹⁵Sherman to Hengelmuller, 20 January 1898, *Papers Relating to the Foreign Relations of the United States,* 78.

⁹⁶Hengelmuller to Sherman, 24 January 1898, *Papers Relating to the Foreign Relations of the United States,* 79.

⁹⁷*The Record of the Times* (Wilkes-Barre), 4 February 1898.

⁹⁸Kulp, ed., *Luzerne Legal Register Reports,* Vol. IX, 77–78.

⁹⁹Henry Hoyt, Assistant Attorney-General of the United States, to William Day, Secretary of State, 8 April 1898, *Papers Relating to the Foreign Relations of the United States,* 87. Governor Henry Martin Hoyt initially appointed Stanley Woodward a judge for the Luzerne County Court in 1897. H. C. Bradsby, ed., *History of Luzerne County, Pennsylvania, with Biographical Selections* (Chicago, 1893), 1482. During the trial Judge Woodward received a number of anonymously written death threats. *The New York Times,* 14 March 1898.

¹⁰⁰Hoyt to Day, 19 April 1898, *Papers Relating to the Foreign Relations of the United States,* 102.

[101] Hoyt to Day, 18 April 1898, *Papers Relating to Foreign Relations of the United States,* 87.

[102] *Ibid.*

[103] Hengelmuller to Day, 26 April 1898, *Papers Relating to Foreign Relations of the United States,* 110.

[104] Robert D. Coxe, Counsel for the Imperial and Royal Austro-Hungarian Government, to Department of State, 20 June 1898, *Papers Relating to Foreign Relations of the United States,* 144.

[105] *Ibid.,* 117.

[106] *Ibid.,* 147.

[107] Charlemagne Tower, United States Minister to Austro-Hungarian Government, to Day, 1 June 1898, *Papers Relating to Foreign Relations of the United States,* 111–117.

[108] John Hay, Secretary of State, to Baron Riedl, Austro-Hungarian Minister to the United States, 4 February 1899, *Papers Relating to Foreign Relations of the United States,* 152–156.

[109] *Ibid.,* 155.

[110] Baron von Riedenau, Austro-Hungarian Minister to the United States, to Hay, 28 April 1899, *Papers Relating to Foreign Relations of the United States,* 38–39.

[111] Calvin D. Davis, *The United States and the First Hague Peace Conference* (Ithaca, 1962), 137–145.

[112] Baron von Riedenau to Hay, 28 April 1899, *Papers Relating to Foreign Relations of the United States,* 39.

[113] *The New York Times,* 11 July 1899.

Chapter Eleven

"DO YOUR DUTY!"
EDITORIAL RESPONSE TO THE ANTHRACITE STRIKE OF 1902

Harold W. Aurand

The anthracite strike of 1902 is interpreted best as a political event. It involved economic issues, such as a minimum wage scale and a twenty per cent increase in contract rates. But the primary goal of the United Mine Workers of America was recognition from the coal companies.[1] The mine operators, on the other hand, sought the destruction of the union within the hard-coal fields. Both objectives were political.

The timing of the strike, during an election year, insured its politicalization within the larger community. The Democratic party attempted to blame the struggle upon the economic policies of the Republicans. Leaders of the GOP sought to identify themselves with an early resolution of the conflict.[2]

The strike did not end with victory on the economic battlefield. Neither belligerent exhausted its ability to continue the struggle by October 1902. Mine workers were unilaterally returning to work in increasing numbers by that time. But movement did not represent a wholesale desertion of the strike.[3] Although the coal companies suffered a sharp decline in revenues during the conflict, they retained ownership of the coal and could easily recoup the losses once production resumed. Wall Street was not apprehensive over the financial implications of the strike. The price of stock of the major anthracite producers remained fairly stable during the conflict.[4]

Third-party political intervention ended the strike in October 1902. The fact that the President of the United States, Theodore Roosevelt, intervened in the strike a month before an election attests to its importance as a political event.

Public reaction transformed a local battle over union recognition into a political issue of national concern. Indeed, it is possible to portray the strike as a trial before the court of public opinion in which the mine owners were found guilty.[5] The portrayal is accurate. But it raises an important question, "Guilty of what?"

Several possible indictments—arrogance of power, inhumane exploitation, and monopoly—quickly come to mind. One immediately thinks of George F. Baer's arrogance in claiming a divine right to administer the property of others. Or, of his contempt for public officials, which caused Benjamin B. Odell, Jr. to explode: "What do you mean by politicians? I want you and all the operators to understand that I am the governor of New York, the chosen representative of seven million people. . . ."[6]

The Lattimer Massacre and the strike of 1900 stimulated a number of publications on the social conditions in the anthracite region. The reading public was thusly well acquainted with the evils of child labor, low wages, underemployment, company housing, and company stores.[7] The ability of the "anthracite combine," a cartel of the major carrier-producers, to fix coal prices was common knowledge.[8]

A survey of major metropolitan newspaper editorials, reveals that none of these issues was critical in the formulation of the editors' response to the strike. Rather, they condemned the operators for dereliction of duty. Both the charge and the politicalization of the strike can be understood only when placed within the context of the industry's market and the ability of the United Mine Workers to disrupt that market.

Anthracite was the most convenient domestic space-heating fuel on the market. It required only a natural draft to provide an even and intense heat. Once ignited, a hard-coal fire could be maintained throughout the heating season with a minimum amount of skill and attention. Moreover, it burned cleanly; it did not form tarry residues in the fire box and emitted little or no smoke.[9] "Smokeless anthracite" was especially attractive in the crowded metropolitan areas of the northeastern United States.[10] Clean and efficient, anthracite was so widely used to heat the homes of America that it was considered "almost as necessary as the ordinary food of life."[11]

Geology confined the production of this "prime neccessity of life"[12] to a compact area in northeastern Pennsylvania.[13] Geographically concentrated, the industry was less cumbersome for labor to organize than its sprawling bituminous counterpart. Ethnic and geo-economic rivalries within the small area, however, presented obstacles to the establishment of a meaningful union, which earlier organizational drives failed to hurdle.[14] But the United Mine Workers demonstrated that it had developed a sense of unity in the area during the 1900 strike. In short, in 1902 the UMW was recognized as having the power to completely shut down the anthracite industry.

Once called, the strike could not be broken with imported labor. In 1889, the Pennsylvania legislature prohibited the employment of non-certified miners in the anthracite industry. To become certified, the candidate had to demonstrate that he had at least two years experience as a miner's laborer in the hard-coal mines and had to pass an examination. The examining board consisted of nine miners each having a minimum of five years experience in the anthracite mines.[15] The certification law gave the miners control over their occupation and insured that as long as they remained loyal to the union not one ton of anthracite would be raised to the surface during the strike.

The distinct possibility of a coal famine alarmed the public. During the pre-strike maneuvers editors grasped at any indication that the conflict would be prevented. On May 1, for example, *The Philadelphia Inquirer* proclaimed that the big strike "will be arbitrated."[16] *The New York Herald* announced that a coal strike was unlikely, but warned its readers that John Mitchell's remarks about a peaceful settlement were guarded.[17] Even as the situation deteriorated, *The New York Times* hopefully commented, "In the meantime, it is gratifying to note certain signs of unusual moderation of temper on both sides which will make an ultimate understanding more practicable."[18] After reporting that there was little doubt of a strike, *The Philadelphia Inquirer* announced on May 7 that the prospects of a strike were minimal.[19] The call for a temporary suspension of work to begin on May 12 destroyed hope that coal production would not be disrupted.[20]

Editorial assessment of responsibility for the strike tended to exonerate labor. John Mitchell received praise for his "caution and moderation" during the pre-strike conferences.[21] His announced willingness to submit the issues to arbitration earned favorable comment.[22]

The editor of *The New York Times,* however, assigned some responsibility for the strike to labor. Mitchell, he argued, placed too much faith in the ability of the National Civic Federation to effect a settlement. In his overconfidence he pursued a policy which left him with no alternative to a strike.[23] In this sense, the strike reflected the internal politics of the United Mine Workers: "The real issue is the fear of the union leaders that unless they do something they will lose prestige and influence with the more turbulent element of their following."[24] But the editor also accused the mine operators of "bad faith" during the pre-strike meetings.[25]

The operators did not show "a serious desire to avoid a strike" during the early conferences.[26] Indeed,

> The attitude of the anthracite mine operators is much to be regretted. From the inception of the trouble they have behaved in a way to encourage the belief and were prepared to do all in their power to bring it about.[27]

The attitude reflected a critical flaw in the character of the operators.

> The mining and transportation of anthracite is not now, and never has been, organized on a sound basis. Its greatest representatives have been little more than opportunists, with whom an immediate advantage has overshadowed everything that looked to the future.[28]

The opportunists desired a strike for it provided them with the chance to destroy the union.[29]

But the operators' war against the union was not a private matter. Rather, it posed "a great public question in which the public has a deep interest and concern."[30] Given that interest, it was absurd to think "that the community should permit the immediate parties to the struggle to conduct it in their own way regardless of the effect upon the country."[31]

For the public, worried about a possible coal famine, it was "not a question of striking, but of ending the strike."[32] The people were entitled to "the uninterrupted production of this universal necessity known as anthracite."[33] Thus, public concern translated into a simple command—produce coal! The demand, give us coal, furnished the textual reference for editorials on the strike. The role of villain fell to the party which at the time, and according to the issue, was perceived as representing the greater obstacle to the resumption of mining.

During the first few weeks of the strike, two issues arose which placed labor in the role of villain. Reports of a possible sympathy strike by soft-coal miners prompted editorial condemnation. The reputed action was denounced as both irresponsible and illegal. Irresponsible in the sense that it increased public anxiety over a coal famine. By so doing, Mitchell was as contemptible as the greedy retail coal dealers who were using the strike as an excuse to gouge the public.[34] A sympathy strike was clearly illegal for it would repudiate existing contracts in the Central Competitive Field.[35]

Mitchell had no intention of violating trade agreements. But he used the fear of a general coal strike to enlist third-party pressure on the operators to reach a settlement. But when that gambit failed, he was obligated to call a national convention of the United Mine Workers to consider the question of including the bituminous coal fields in the strike.[36]

The convention call raised doubts over Mitchell's leadership abilities. Once again it appeared to the editor of *The New York Times* that by "talking too much and saying more than he meant," Mitchell painted himself into a political corner. He would either have to support an irresponsible and illegal program or appear insincere to the anthracite miners whom he promised to aid.[37]

Mitchell extricated himself from the dilemma during a speech before the convention. He advised against the general strike, urged the soft coal miners to financially aid the hard coal strikers, and proposed a set of recommendations concerning that aid. The speech and the convention's acceptance of Mitchell's recommendation earned wide applause for the union.[38]

Editors were less forgiving on the second issue. Indeed, as late as September the Philadelphia *Public Ledger* referred to it as an illustration of "the dense stupidity of Mr. Mitchell."[39] On May 21, the Anthracite Executive Committee ordered firemen, engineers, and pumpmen to discontinue work unless they were granted an eight-hour day without a reduction of wages. Traditionally, these men worked during strikes, for they kept the mines clear of water. Water was a serious problem in the anthracite mines; an average of ten tons of water was lifted for every ton of coal raised to the surface.[40] Editors perceived the order as an effort to flood the mines.

They quickly denounced it as "wholly unlike the generally sagacious and conservative policy hitherto pursued by Mr. Mitchell."[41] The order clearly violated their demand for an immediate resumption of production. At best, flooding the mines would prohibit mining for weeks or even months after the cessation of hostilities. At worst, the flooded mines would never be reopened. The editors supported the rights of property; the operators, they warned labor, had the right and the responsibility to protect the mines.[42]

Since firemen, engineers, and pumpmen were not protected by certification, the operators quickly moved to replace the strikers. Employment agencies in New York, Philadelphia, and Baltimore recruited men for work in the hard-coal fields.[43] The coal companies increased their private police forces to protect the strike-breakers; within one week the Governor of Pennsylvania commissioned 1,170 new coal and iron policemen.[44] Under such heavy protection, the operators managed to keep most of the mines clear of water.[45]

Labor responded to the build-up of the private police forces with disparaging comments. Union spokesmen alleged that some of the new recruits were criminals. Indeed it was reported that the UMW hired the Pinkerton Detective Agency to investigate the background of some of the new policemen.[46] Regarding the coal and iron police as protectors of property, editors dismissed labor's criticism as "ill-advised talk."[47] Indeed, "the lack of confidence in President Mitchell's honesty" stemmed from "his efforts to make the Coal and Iron Police odious."[48]

The apparent ease with which the coal companies replaced the striking pumpmen created a supposition among editors that the strike's outcome had been decided. The theory provided the backdrop against which they could blame both labor and management for needlessly continuing the conflict. One scenario, supported by the operators' periodic announcements of men returning to work and predictions of an early disintegration of the strike, was that labor lost the war. In a real sense, it was not even a contest.

> From the first day of the struggle to this day all the odds have been against the strikers. They were outclassed in intelligence, shrewdness, organization, and leadership. They had no tactical, sagacious, trained captains like President Baer to lead them.[49]

Yet the strike persisted in violation of the editors' demand for an immediate resumption of mining.[50]

Editors placed the responsibility for labor's failure to recognize the inevitable upon John Mitchell. In July, according to *The New York Herald,* his friends advised him that the strike was doomed and recommended that he end the struggle before the union was destroyed.[51] But his pride prevented him from acknowledging his blunder in calling the strike.[52] Mitchell's failure to accept the obvious raised the question of his capacity for leadership; good leaders, after all, have "the power to recognize defeat and the courage to retreat when nothing else could be gained by continued resistance."[53]

The same scenario, however, could be used to demonstrate the guilt of the operators for continuing the strike. One argument was that they lacked "the courage of self-assertion."[54] They defeated labor, but were too timid to consolidate their victory by reopening the mines. Instead, they idly talked of reopening. "Now as always," *The Philadelphia Inquirer*, complained "the way to resume work is to resume, not talk about it."[55] By August the operators managed to reopen fourteen mines. But the effort was denounced as a mere pretense.[56] Editors would accept nothing less than a "well directed and sustained effort to recover control of the properties of which they were deprived when the union stopped mining."[57] Regaining control seemed to be a simple task; the operators could double their guards, demand increased military protection, evict the strikers, and reopen the mines with new employees. But a wiser policy, they were advised, would be to offer some concessions to the strikers.[58]

As superior men, the operators were obligated to be magnanimous in victory. They could end the strike by offering such fair, or even generous, terms that the miners could accept defeat without humiliation.[59] Or, they could accept Mitchell's repeated offers to settle the strike by arbitration. Most editors urged arbitration. But the operators refused to expedite the conclusion of the strike.

Refusal to grant concessions or accept arbitration created the suspicion that the operators prolonged the strike simply to punish their men.[60] By so doing they were violating the public's interest. "The coal barons tax the whole community by their contumacious refusal to listen to any terms for the settlement of the strike which are not dictated by themselves."[61]

The operators' inability to impose their will upon the strikers caused some to suggest that management lost the war. "The impressive feature seems to us to be that the operator knows he must have the present miners to work for him. There is no effort to start the mines with outsiders."[62] If such was the case, the operators must quickly come to terms with the union and resume mining. Unfortunately, they turned a deaf ear to the editors' demand that they either open the mines without the union or submit to arbitration.[63]

Frustrated by the apparent unwillingness of either side to end the strike, the public wished a "pox on both."[64] But the curse had little effect. More direct action appeared necessary and several radical solutions to the problem were advanced. It was proposed that the courts enjoin the operators to reopen the mines.[65] The New York Democratic party called for the nationalization of the industry.[66]

The Republicans correctly gauged the political implications of the growing sentiment against the strike. Aware that during hard times the vote goes against the "ins," they attempted to associate themselves with an early resolution of the conflict. Marcus Hanna, of course, was closely identified with the National Civic Federation's efforts to mediate a compromise. During the strike Senators Mathew Quay, Thomas Platt, and Boise Penrose, among others, held widely publicized conferences with the belligerents in an attempt to effect a settlement.[67] The strike became embroiled in a power struggle within the Pennsylvania GOP as the Quay-Penrose and the Stone-Elkins-Finn factions both competed for the glory of ending the strike.[68]

The response to political intervention was negative. Mine operators blamed the politicos for prolonging the strike.[69] Editors, perhaps concerned by the antics of "machine politicians," such as Matt Quay, warned "meddlers and politicians" to leave the issue alone.[70]

The warning, however, did not extend to the President of the United States. In June, the New York Board of Trade and Transportation asked Theodore Roosevelt to arbitrate the strike.[71] Although he denied the Board of Trade's request, Roosevelt instructed Carroll W. Wright, Commissioner of Labor, to investigate the struggle. Published in September, Wright's report carried a disclaimer of the President's constitutional power to act in the matter.[72]

By late September, mounting political pressure convinced Roosevelt that he must do something about the strike. The historic conference at the temporary White House on October 3, unfortunately, failed to produce a settlement. The apparent failure of the conference increased public frustration. "When the president fails," *The Philadelphia Inquirer* lamented, "it is useless for others to try" to end the strike.[74]

Only the operators were in a position to end the conflict and editors insisted that they had a public obligation to do so.

> A corporation that mines coal is a public concern.
>
> The coal companies which have refused to end the strike have no right to consider the question of the suspension of mining as a private one. It is not. It is a public one. Coal is almost as necessary as the ordinary food of life.
>
> If they can furnish it without employing the strikers, let them do it. But with or without the strikers we must have coal.[75]

The mine owners' failure to supply the public with coal was nothing less than an act of "non-feasance" and a "warrant for the withdrawal of privileges."[76]

In their defense, operators argued that fear prevented the miners from returning to work.[77] The use of terror to prolong the strike became the most damning charge against labor. Ironically, the first person killed in the conflict, twelve-year-old Charles McCann, was the victim of the coal and iron police.[78] But reports of strikers engaged in violent activities became so numerous during the first month of the strike that one editor concluded:

> The laws of Pennsylvania have been for weeks trampled on and set to naught by the striking miners. A reign of terror inspired by violence and menace has there surplanted the reign of law.[79]

In late July, a mob in Shenandoah added substance to the charge of labor-inspired terror by beating Joseph Bedacock to death. As workers began returning to work during August and September, the level of violence dramatically increased.[80]

Editors held Mitchell accountable for the "prevailing state of anarchy" in the coal regions:

> He has allowed his sympathy with the anthracite miners to make him conveniently blind to the efforts of his followers to maintain a reign of terror in the coal regions and has talked like a demagogue whenever he has deemed it wise to speak at all. This is not greatness. The Moses of the labor movement is not Mitchell.[81]

Some noted that Mitchell advised his followers to obey the law, but charged him with not being "forceful enough" when admonishing the strikers to restrain from violence.[82]

The mine operators felt secure in raising the issue of labor violence during the White House conference. "Mitchell," Baer told Roosevelt, "must stop his people from killing, maiming, and abusing Pennsylvania citizens and from destroying property."[83] Much of the press agreed. The *New York Times* dismissed Mitchell's reply to Baer's remarks as "a shifty and dishonest evasion of the truth and a repetition of the silly falsehood" that the coal and iron police generated much of the violence in the region.[84] The Philadelphia *Public Ledger* wondered if Roosevelt had not demeaned the presidency by meeting with the "chief executive of the organized hosts of lawlessness, riot, and misrule."[85]

Labor violence smacked of hypocrisy. It was an attempt to deny other men the same right—the freedom to work on terms acceptable to themselves—which the strikers claimed. More importantly, it violated public interest. One editor informed labor that the coal companies "owe it to the public to work the mines with new men, if necessary" and warned "the new men will be protected."[86]

Final responsibility for protection of strike breakers fell upon government. Governor William A. Stone was, at first, reluctant to send the National Guard into the coal fields. The riot in Shenandoah, however, caused him to order troops into that city. He responded to the growing violence during August and September by dispatching soldiers to all of the anthracite counties. Yet editors denounced Stone's "timid discretion" and accused him of inadequately garrisoning the region.[87] On October 6, Stone ordered the entire National Guard into the area.

Stone's action placed the burden of proof upon the mine owners: "The operators will have a chance to demonstrate their claim that once freed from the terrorism of the strikers enough coal can be produced to relieve the famine."[88] But the presence of over eight thousand soldiers failed to end the strike. Indeed, the men voted unanimously to continue the struggle.[89] The decision to maintain the strike destroyed the operators' defense that terror prevented them from reopening the mines.

> All of the troops in Pennsylvania have been called out. It is the claim of the operators that protection alone was required. As a matter of fact, coal is not being mined, although full protection has been granted. The people require and demand coal. It is the absolute duty of the companies to give us coal.[90]

The operators not only failed to fulfill their duty, they repudiated the obligation by insisting that "the management of business belongs to the owners." Their "public be damned" attitude reflected unpardonable arrogance.

> A body of men who, in the United States and in the twentieth century disclaim a clear moral responsibility in the administration of a monopoly of one of our most valuable natural resources, because it has not yet been made a legal responsibility, display great temerity wholly untempered by discretion.[91]

From the mine operators' perspective, however, the editors articulated a totally unfair policy. They demanded that the coal companies either do the impossible or surrender to labor without a fight. It must be remembered that the key issue in the strike was the status of the United Mine Workers. To grant concessions to the strikers would only encourage them to maintain their organization. To accept arbitration implied recognition of the union. Both possibilities could only be interpreted as a defeat for the coal companies.

The Certification Law of 1889 precluded the replacing of the striking miners with outsiders and, therefore, prevented an immediate resumption of production. Moreover, the editors' demand was counterproductive. It encouraged the men to continue the struggle in anticipation of gaining in the political arena what they could not hope to win on the economic battlefield.

The operators responded to the strike of 1902 as if it were a private contest. Denied the option of a *blitzkrieg* by miners' certification, they fought a war of attrition. Indeed, they were candid about their strategy; Robert M. Olyphant, president of the Delaware and Hudson Company, explained, for example:

> After our men have been idle for a while, with their families to support and no wages coming in, they may take a different view of the strike. We are confident that they will regret their action and be glad to resume work on the old terms.[92]

Replacing the striking pumpmen was the decisive battle in the operators' campaign. Assured of the preservation of their property, they could wait until the necessities of life compelled their employees to return to work on the companies' terms. Past experience suggested that it required six to seven months to starve the mine workers into submission.[93] And the number of men returning to work during September indicated that the strategy was sound. Once the men returned to work on the companies' terms, the demand for coal would be easily met.

Unfortunately for the operators, editors, alarmed by the possible social repercussions of a coal famine, refused to patiently sit out a war of attrition. Instead, they charged the coal companies with an obligation to the public to immediately resume mining with or without the union.

NOTES

[1] Robert J. Cornell, *The Anthracite Strike of 1902* (New York, 1957), 90.
[2] *The New York Times,* 20 August 1902, 9; 5 September 1902, 8.
[3] Philadelphia *Public Ledger,* 24 September 1902, 8.
[4] Stock prices varied from day to day. But during October they were near or above the closing prices of the first day of the strike.
[5] Robert L. Reynolds, "The Coal Kings Come to Judgement," *American Heritage,* XI (April 1960), 54–61, 94–100.
[6] *The New York Herald,* 11 October 1902, 4.
[7] See, for example, Henry Edward Rood, "A Pennsylvania Colliery Village," *The Century Magazine,* LV (April 1898), 809–821; Jay Hambridge, "An Artist's Impression of the Colliery Region," *The Century Magazine,* LV (April 1898), 822–828; Charles B. Spahr, "America's Working People: The Coal Miners of Pennsylvania," *The Outlook,* LXIII (1899), 805–812.

[8] Several official investigations of the anthracite cartel occurred during the last quarter of the nineteenth century.
[9] E. Willard Miller, "The Southern Anthracite Region: A Problem Area," *Economic Geography,* XXXI (October 1955), 331–332.
[10] Concern for clean air caused New York City to prohibit the burning of bituminous coal. (*The New York Herald* 17 June 1902, 3.)
[11] *The Philadelphia Inquirer,* 16 September 1902, 8.
[12] *The New York Herald,* 12 June 1902, 8.
[13] All of the anthracite deposits are found within a 439 square mile area.
[14] Neither the Workingmen's Benevolent Association nor the "Joint Committee" was able to effect an industry-wide strike. See Harold W. Aurand, *From the Molly Maguires to the United Mine Workers* (Philadelphia, 1971).
[15] *Pennsylvania Laws,* 1889.
[16] 1 May 1902, 1.
[17] 1 May 1902, 3.
[18] 5 May 1902, 8.
[19] 6 May 1902, 1; 7 May 1902, 1.
[20] The suspension was made permanent on 14 May 1902.
[21] Philadelphia *Public Ledger,* 12 May 1902, 8.
[22] *Ibid.,* 1 October 1902, 8; *The Philadelphia Inquirer,* 25 June 1902, 8.
[23] *The New York Times,* 16 May 1902, 8.
[24] *Ibid.,* 19 May 1902, 8.
[25] *Ibid.,* 16 May 1902, 8.
[26] *The Philadelphia Inquirer,* 18 May 1902, 8.
[27] *The New York Times,* 6 May 1902, 8.
[28] *Ibid.,* 8 June 1902, 2.
[29] *The New York Herald,* 27 August 1902, 3; 12 May 1902, 3.
[30] Philadelphia *Public Ledger,* 24 July 1902, 4.
[31] *The New York Herald,* 18 June 1902, 8.
[32] *The Philadelphia Inquirer,* 21 May 1902, 8.
[33] *The New York Herald,* 30 May 1902, 8.
[34] In 1898, the United Mine Workers agreed to a joint conference with the middle competitive field operators which provided for annual agreements.
[35] Cornell, *op. cit.,* 100–119.
[36] *The New York Herald,* 17 May 1902, 8.
[37] 19 June 1902, 8.
[38] *The New York Herald,* 18 July 1902, 6.
[39] 12 September 1902, 8.
[40] Cornell, *op, cit.,* 98–99.
[41] Philadelphia *Public Ledger,* 23 May 1902, 8.
[42] *Ibid.,* 5 June 1902.
[43] *The New York Times,* 3 June 1902, 3; 4 June 1902, 3.
[44] *Ibid.,* 1 June 1902, 1.
[45] Cornell, *op. cit.,* 99.
[46] *The New York Times,* 2 June 1902, 4.
[47] *Ibid.,* 4 June 1902, 4.
[48] *Ibid.,* 5 June 1902, 1.
[49] Philadelphia *Public Ledger,* 26 September 1902, 8.
[50] *Ibid.,* 2 October 1902, 8. *The New York Times,* 16 September 1902, 8.
[51] *The New York Herald,* 8 July 1902, 6.

[52] Philadelphia *Public Ledger,* 17 September 1902, 8.
[53] *The New York Times,* 22 September 1902, 8.
[54] *Ibid.,* 1 July 1902, 8.
[55] 11 August 1902, 8.
[56] *The New York Times,* 29 August 1902, 8.
[57] *Ibid.,* 4 September 1902, 2.
[58] *Ibid.,* 13 September 1902, 8.
[59] Philadelphia *Public Ledger,* 14 October 1902, 8.
[60] *The New York Times,* 8 July 1902, 8.
[61] Philadelphia *Public Ledger,* 24 July 1902, 4.
[62] *The Philadelphia Inquirer,* 29 July 1902, 8.
[63] *Ibid.,* 6 August 1902, 8. *The New York Herald,* 3 October 1902, 10.
[64] *The New York Herald,* 9 October 1902, 10.
[65] *The New York Herald,* 22 September 1902, 8; 12 October 1902, 8. *The Philadelphia Inquirer,* 9 August 1902, 8.
[66] Philadelphia *Public Ledger,* 3 October 1902, 8.
[67] *The New York Herald,* 11 October 1902, 4.
[68] Philadelphia *Public Ledger,* 11 September 1902, 8.
[69] *The New York Herald,* 12 October 1902, 4.
[70] Philadelphia *Public Ledger,* 9 September 1902, 8.
[71] *The New York Times,* 5 June 1902, 3.
[72] Cornell, *op. cit.,* 105–110.
[73] *Ibid.,* 172–189.
[74] 5 October 1902, 8.
[75] *The Philadelphia Inquirer,* 2 August 1902.
[76] *The New York Times,* 11 August 1902, 6.
[77] *Ibid.,* 2 October 1902, 8.
[78] *Ibid.,* 8 June 1902, 2.
[79] *Ibid.,* 21 June 1902, 8.
[80] Cornell, *op. cit.,* 151–152.
[81] *The New York Times,* 19 July 1902, 8.
[82] Philadelphia *Public Ledger,* 15 September 1902, 8.
[83] Quoted by Cornell, *op. cit.* 184.
[84] *The New York Times,* 6 October 1902, 8.
[85] *Ibid.,* 10 October 1902, 8.
[86] *The New York Herald,* 5 August 1902, 8.
[87] *The New York Times,* 8 October 1902, 8; Philadelphia *Public Ledger,* 5 October 1902, 8.
[88] *The New York Herald,* 7 October 1902, 6.
[89] Cornell, *op. cit.,* 184.
[90] *The Philadelphia Inquirer,* 11 October 1902, 8.
[91] *The New York Times,* 14 August 1902, 8.
[92] Philadelphia *Public Ledger,* 16 May 1902, 1.
[93] Both the strike of 1875 and the strike of 1887–88 lasted slightly longer than six months.

Chapter Twelve

COMMENTARY: ETHNIC RESPONSES TO THE LATTIMER MASSACRE AND "DO YOUR DUTY"

Ronald L. Filippelli

These two papers share several admirable characteristics. Both are straightforward and modestly conceived, ideal qualities for a conference such as this. Both are well-researched and thought-provoking. That is to say that both authors did what they set out to do, and did it well.

Although each paper is about a response to a labor dispute, the similarities are few. This increases the difficulty for the commentator. In such circumstances historians frequently feel the need to strive for synthesis, even when the effort requires the most outrageous contortions. I shall resist the temptation and take the papers one at a time.

Professor Turner provides us with a look at a little-studied facet of American social history, the response of immigrants, in this case the Hazleton Slavic community, to an act of violence against them. The fact that the "Lattimer Massacre" has gone largely unnoticed by historians is in itself a reflection of the low esteem in which the victims were held by their contemporaries. Given their insecurity in a strange land, any organized reaction on the part of the Slavic community deserves credit. What I would like to explore a bit more than Professor Turner are some implications behind the nature of that response.

The response, in spite of the intensity of the outrage, was fundamentally conservative, as was the behavior of the miners in the original labor dispute. Not even the murder of their countrymen by law officers shook the immigrants' need to believe in the basic justice of the system. In seeking legal redress for the wrong done to them, they petitioned their government peacefully; they followed the moderate leadership of businessmen, a labor leader, and a priest; and they even suffered silently the supercilious lectures of the local newspaper. Feeling isolated and besieged, they adopted a course of action that, in theory, was much more suitable to the benevolent autocracy of their homeland, the Austro-Hungarian Empire, than to the ostensibly just and democratic society to which they had immigrated.

Indeed, so cautious was the controlled response that after the slaughter one was hard-pressed to find, at least in Professor Turner's paper, any reference to the fact that the events at Lattimer had, after all, resulted from a labor dispute between immigrant coal miners seeking redress of their grievances and the coal operators. Had this been the case only in the local press, it would be understandable, dependent as it usually was on the mining industry. But the degree to which the Slavic community and its national committee were silent on the real issue is striking. Their conservative behavior at least signified that they were far from the dangerous mob that Sheriff Martin had felt compelled to fire into in order to preserve public order. At worst, it symbolized the degree to which the immigrant workers had either been cowed by fear or co-opted by their moderate leaders—leaders with a large stake in upholding the prevailing value system. Whatever else it may be, the story of the Slavic community's feeble and unsuccessful response to the Lattimer massacre is a lesson in social control.

Professor Aurand looks to another quarter for responses to the Anthracite Strike of 1902. Aurand focuses on the editors of the major Philadelphia and New York newspapers. His concern is not with the day-to-day coverage, but with the editorial comment. This is an important distinction. News coverage of the strike reached a wide audience. It chronicled the violence, the hardship, and the drama of the epic confrontation between labor and capital. The editors wrote for a different audience, one made up of elites, including the coal operators.

In 1902, eastern newspaper editors, as much as any other occupational category, would have fallen under the broad classification historians have labeled Progressive, and the anthracite strike was to become, in large measure because of their role, the archtypical Progressive-era strike.

Progressivism was far from a homogeneous movement. Historians have located so many major streams and minor tributaries flowing into the river of progressivism that one must use extreme caution in trying to identify factors that held the diverse reform movement together.

The progressives shared a concern with the negative impact of industrial capitalism. They understood that they were dealing with a modernizing society. Many of the major ideas that contributed to progressivism—scientific management, industrial efficiency, industrial democracy—were modernizing ideas. The progressives also shared a preoccupation with the role of modern economic institutions because they believed that economics and economic relationships largely determined social maladjustment of progress. Their preoccupation with economic determinism led them to define society in terms of its functional and productive elements. Thus the roles of managers, owners, workers, and consumers within the economic system fascinated them.

The consequence they feared the most from industrial capitalism was class conflict. They focused much of their attention, through their support for arbitration, scientific management, and welfare capitalism schemes, on avoiding Marx's model of inevitable conflict. Because of the power of big business and its resistance to reform, they assigned government a major role in achieving a prosperous, harmonious, and just society.

Although the progressives supported reform at the local level to break the power of the special interests, they assigned the main role to the national government. What they had in mind was a government of professionals, almost a neutral and efficient technocracy. While they believed that positive state action was necessary to manage an industrial economy, they had a disdain for the rough and tumble and unpredictability of partisan politics.

More than anything else the progressives feared class struggle, and they blamed it on the excesses of social Darwinism and extreme economic individualism—an economic individualism that resulted in the sacrifice of the public interest. The great labor wars of the late nineteenth and early twentieth centuries confirmed their worst fears. It was in this context that the editors viewed the Anthracite Strike of 1902, and assigned the major responsibility for it to the mine owners.

It is interesting, and entirely consistent, as Professor Aurand demonstrates, that the editors did not ground their indictment of the coal operators on inhumane exploitation and arrogance of power. Nor did they spend much time on the immediate economic and representational issues that had caused the strike. These matters could better be left to the news columns. The editors had bigger fish to fry. The issue that interested them was the responsible management of a public utility. As Aurand perceptively stresses, the indictment of the operators was based on their incompetence.

This is understandable coming from a small group of influential editors who valued professionalism, objectivity, and efficiency. Implicit in their criticism lay the idea that the men who owned the mines were stewards of the public interest. This was a long way from the traditional laissez faire conception of the social benefits of unrestrained individualism. Frequently, the editors emphasized the public character of the coal industry and of the strike.

"A corporation that mines coal is a public concern," wrote the editor of the *Philadelphia Inquirer*. "The coal companies which have refused to end the strike have no right to consider the question of the suspension of mining a private one . . . coal is almost as necessary as the ordinary food of life." Nor did the rights of the strikers supercede the public interest. "If they [the operators] can furnish it without employing the strikers, let them do it," he continued. "But with or without the strikers, we must have coal."

Failure of the coal operators to assume this public responsibility, according to the editor of the *Philadelphia Ledger,* was the result of their "contumacious refusal to listen to any terms for the settlement of the strike which are not dictated by themselves." Here was the heart of the matter. The problem was not that the coal operators possessed too much power by virtue of private ownership of a public utility, but that they persisted in acting like the laissez faire capitalists they were. It would be difficult to find a clearer example of a conflict between two elites. An angry *New York Times* editor summed up the classic progressive attack on the coal operators:

> The mining and transportation of anthracite, is not now, and never has been, organized on a sound basis. Its greatest representatives have been little more than opportunists, with whom an immediate advantage has overshadowed everything that looked to the future.

When private advantage so overshadowed the public welfare, the editors turned naturally to government intervention by state and local officials whom the editor of the *Philadelphia Public Ledger* damned as "meddlers and politicians." Few words in the progressive lexicon carried as much reprobation as "politician." Indeed, elected Pennsylvania officials such as Boise Penrose, Matthew Quay, and Thomas Platt, who tried to end the strike to political advantage, represented the worst examples of the self-interest, partisan politics, corruption, and inefficiency that had given a major impetus to progressivism in the first place.

Although the actual strike was limited to northeastern Pennsylvania, the editors always viewed it as a national question. Their call for intervention by the national government reflected this. It also reflected the progressive faith in detached and objective solutions. Theodore Roosevelt's appointment of the Anthracite Coal Strike Commission—arbitration by "experts" after a thorough study of the "facts"—was a prototypical progressive solution. Naturally, the operators, who were winning the strike, had to be forced to arbitration. Roosevelt railed against them for increasing "the burden" on those who stood between the people and the danger of the spread of socialism. Undoubtedly the editors would have included themselves in that select company.

Chapter Thirteen

THE ETHNO-RELIGIOUS FACTOR REACHES FRUITION:

THE POLITICS OF HARD COAL, 1945–1972

William A. Gudelunas

The role of ethnicity and religion in influencing national political behavior has been a subject of conflicting opinions for some time. Some analysts argue that religion, race, and ethnicity would fade away as political determinants, to be replaced by functional economic interests[1] or the recognition of one's true class interests.[2] Others argue that ethnicity will decline, but religion will remain salient for quite sometime. For example, Glazer and Moynihan see religion and race as becoming the major defining categories for groups, even as the nationality of immigrant groups declines in importance.[3] Therefore, four main groups have evolved, white Protestants, Catholics, Jews, and Negroes. Still others see ethnicity and religion as remaining important, and/or becoming once again, as socially defining and politically influential categories.[4]

Evidence on the subject also is conflicting. To cite but two examples, Greeley examined a national sample and found distinct differences in the political participation of American ethnic groups. On the other hand, Knoke and Felson also used a national sample but found one's ethnic group having very little to do with one's party affiliation. While religion was found to be slightly more important, both characteristics were found to be of declining importance politically.[5]

The authors of both of the articles suggest that follow-up studies at local levels would provide good supplements to their work. Much of the evidence gathered through use of national samples finds the number of members of certain ethno-religious groups to be very small. Moreover, local and regional differences are obscured, overgeneralization is all too easy, and variations in political behavior due to the specific issues are sacrificed for sometimes poorly defined "general trends." Micro-level analyses of local geographic areas are badly needed to supplement and correct conclusions reached solely on the basis of macro-level, nationwide research. In addition, the history of the ethno-religious groups and their changing place in the social structures of specific geographic areas must be taken into consideration.

This paper attempts to assess the role of ethno-religious factors in the political behavior of a specific region, Pennsylvania's lower anthracite area, between the end of World War II and the early 1970s. This will lead, it is hoped, to a more clearly defined understanding of the contemporary and historic impact of ethno-religious politics.[6]

Pennsylvania's lower anthracite field (Schuylkill County) was historically the scene of violent ethnic conflicts, politically and otherwise. Schuylkill County has become synonymous, for example, with the "Molly Maguire" episodes. Many studies have noted the fact that the county's politics were firmly rooted in ethno-religious concerns.[7] This ethno-religious political orientation lingered well into the twentieth century.[8]

In the nineteenth century, the county was dominated by the Democrats. This was caused by an alliance between Irish Catholics and German Protestants.[9] The county's Germans followed a national trend and abandoned the Democrats in the latter part of the nineteenth century; this isolated the Irish politically and ushered in a Republican control of the county which has never ended. There was a great influx of eastern Europeans into the county between 1880 and 1920, but these new "ethnic" voters (mainly Lithuanians and Poles) tended to divide their votes and did not significantly alter the Republican orientation of the county.[10] In short, twentieth-century Schuylkill County was politically dominated by the Republicans. The Republican based their strength on support given them by the county's Protestants (mainly German with considerable Welsh and English factions). The twentieth-century county Democrats could rely on solid support only from the Irish Catholics. The area's large east European Catholic community tended to split nearly evenly between the parties, allowing Republican dominance to continue.

Would this ethno-religious political orientation continue after World War II? The voters would come to represent later generations of "immigrants." The county would be forced to endure economic changes which could be fairly labeled "cataclysmic." There would be a marked loss of population. Certainly, factors would present themselves between World War II and the early 1970s which could have led to a decline in ethno-religious political behavior. Yet, the county's patterns of ethno-religious voting were so deeply ingrained that they could well resist pressure to disintegrate them.

Schuylkill County's anthracite industry suffered its demise after World War II (especially in the early 1950s). This amounted to a virtual economic revolution within the county. The region's entire economic base shifted within one turbulent decade. The "anthracite decline" of the early 1950s virtually paralyzed the area economically. In 1953, 15,116 people were employed in the county's anthracite industry. This figure dropped to 10,123 workers in 1955 and to an unbelievable 6,900 by 1960.[11] The following chart clearly shows the anthracite industry's decline in the county after World War II.

TABLE I[12]

THE DECLINE OF KING COAL IN SCHUYLKILL COUNTY

Year	Tons Mined	Anthracite Employees
1945	16,731,000	19,000
1950	14,066,000	12,000
1954	10,142,000	8,500
1960	6,933,000	6,900
1968	4,297,000	2,646

The county had been saved from complete oblivion only by the growth of the apparel industry. By 1960, "manufacturing" was the primary employer in the county, while "mining" employment had slipped to a distant third place.[13] The Bureau of Employment Security habitually listed Schuylkill as an area of "substantial and persistant" unemployment. The official unemployment rate in the county was 16.1 per cent in the summer of 1960.[14] By 1960, the per capita income in the county was listed at $1,910, compared to the State average of $2,450.[15]

In brief, the Schuylkill County of the 1960s and 1970s was economically much different than the county in the immediate post-World War II era. Anthracite was no longer "King," low-paying garment industry-type jobs were now more prevalent than higher-paying mine jobs. Unemployment consistently ran well above state and national figures. Economic stagnation typified the area. Certainly, economic turmoil of this magnitude could threaten the established political order.

Ethnically, the county was still quite diverse, consisting of four basic groups: German Protestants, English and Welsh Protestants, Irish Catholics, and eastern European Catholics. The Germans, Lutheran or Reformed, lived mainly in the farm areas of the county and had little connection with the mining industry. Most of the county's Germans could trace their American ancestry to the pre-Revolutionary period, having migrated to farms which constituted the northern extension of the "Pennsylvania Dutch" area. Their social institutions and culture have retained more that is common with their fellow Germans to the south than with their more proximate fellow countians. Indeed, the German districts of the county were traditionally typified by their homogeneity. This characteristic still prevailed after 1945. The German agricultural areas of Schuylkill County simply never attracted a significant number of non-Germans. The German "boroughs" and "valleys" in the 1960s and 1970s were virtually as pure German as their counterparts of the nineteenth century. It was almost as easy to find "German voting districts" in 1970 as it was in 1870.

Most English and Welsh Protestants of the area can trace their roots to the original anthracite boom of the 1840s. The English and the Welsh of the mid-nineteenth century possessed the mining skills which were required to exploit the hard-coal veins of the Schuylkill Region. Because of their mining orientation, the vast majority of these Protestants settled in the urban or semi-urbanized areas in the heart of Schuylkill County's anthracite districts. By the mid-twentieth century, the English and Welsh who remained in the area inhabited basically the same localities as did their forebearers of a century earlier.

About half of the inhabitants of the county were Roman Catholics of Irish, Lithuanian, Polish, Italian, and German origins. Of these groups, the Irish, Lithuanians, and Poles were the most numerous.[16] The non-Irish Catholics of the area were generally second- and third-generation Americans. The Irish ranged from second- to fifth-generation Americans. There also were present within the county a small percentage of second- and third-generation Ukrainian, Hungarian, and Russian Catholics who were Eastern Rite rather than Roman. The coinciding of the Irish potato blights of the 1840s and 1850s with an increasing demand for mine labor resulted in the migration of Irish Catholics to the county. The fact that many English and Welsh miners possessed mining skills before arriving in the county, reinforced by ethnic and religious antagonism, resulted in the Irish entering the area at the bottom of the economic and social ladder. They lived in separate areas and developed institutions distinct from their English and Welsh neighbors. The economic, ethnic, and religious differences between these groups combined to produce some of the most turbulent labor conflict in this country's history.

The memory of these conflicts is still very much alive among Schuylkill Countians and continues to reinforce a strong ethnic identification among the county's Irish Americans.

The county's east Europeans (mainly Lithuanian and Polish) were part of the wave of "new immigrants" that flooded the United States between the Civil War and World War I. These people naturally gravitated toward the low-skill, low-paying jobs of the anthracite region. Their inability to speak English caused them to settle in "patches" which were composed of fellow ethnics. Consequently, names like "Polish Hill," "Lithuanian Gardens," and "Hunkie Flats" became well known to most residents of the lower anthracite region.

As late as the early 1970s, many of these east Europeans hung tenaciously to reminders of their heritage. All of the mining areas of the county still had Polish, Lithuanian, and Italian churches rather than territorial "regional" churches. The coal region was still marked by "Polish Clubs," "Lithuanian

TABLE II[17]

MEMBERSHIP IN "ETHNIC" CATHOLIC CHURCHES SCHUYLKILL COUNTY

Ethnic Groups	No. of Churches	1960 Souls—% of County's Population	1970 Souls—% of County's Population
Lithuanian	10	12,318 (7.1%)	11,190 (6.9%)
Polish	8	9,782 (5.6%)	8,885 (5.5%)
Slovakian	5	4,469 (2.5%)	3,637 (2.3%)
German Catholic	5	4,251 (2.4%)	3,660 (2.3%)
Italian	5	3,862 (2.2%)	3,729 (2.3%)

schools," and "Italian fraternities." Many ethnic parochial schools still included instruction of the "native" language in their curricula. The Catholic hierarchy was still careful to assign "Polish" pastors to "Polish" parishes. Virtually all "ethnic" churches still used the native tongue at some masses each Sunday. This strong identification with ethnic heritage continues to personify Schuylkill County into the 1980s.

The continuing strength of this ethnic identification by Schuylkill Countians is exemplified by the support shown in the "ethnic" Catholic churches in the county. The "Spiritual Reports" required annually of the pastors of all Catholic parishes reveal that few parishioners abondoned the ethnic churches in the 1950–1980 era.

There were (are) also eleven Eastern Rite or Byzantine Catholic Churches in the county during this period. All in all, the county could be simply termed slightly more than fifty per cent Catholic throughout this period. The official figures on the county's religious groups in the mid 1960s revealed:

TABLE III[18]

CHURCHES & CHURCH MEMBERSHIPS, SCHUYLKILL COUNTY, 1964–1965

Group	No. of Churches	Membership
Catholic	64	70,937
All Protestant	203	59,041
Lutheran	60	29,350
Methodist	26	5,790
United Church of Christ	57	13,212

(The last three were the groups with the largest number of Protestant churches.)

It must be remembered that the county's religious groups were still somewhat "segregated" into the 1970s. The Catholics lived mainly in the declining coal areas while the Protestants, especially the Germans, tended to inhabit the farm regions to the west and south of the coal belt.[19] These fairly distinctive demographic patterns lend themselves quite well to social scientific analysis.

It also should be noted that the county lost population throughout this period. The population dropped from 200,577 in 1950 to 160,089 in 1970 (−20.1 per cent). However, the statistics indicate that this tremendous out-migration did not alter the area's ethnic composition in an appreciable manner.

There were other changes in population which could conceivably have altered the county's political leanings. The area became notably older and more dependent on government programs, such as social security. By the early 1970s, 26.6 per cent (42,547 people) of the county's population was aged fifty-five or over.[20] Approximately the same percentage of people depended on social security or "black lung" benefits for at least part of their income.[21]

This meant that Schuylkill County was a better than fifty per cent Catholic county, growing older and increasingly dependent upon governmental programs, such as social security and unemployment compensation. This would make the lower anthracite region a "classic" liberal, Democratic area in the minds of most people. When these factors are combined with the fact that the county went through a veritable economic transformation between 1950 and 1970, a political change of mass proportions would not be surprising. However, the voting figures reveal that the county's legendary ethno-religious voting patterns prevailed despite these dramatic changes. The net result would be a Republican ascendancy in the county well into the 1970s.

The twentieth-century dominance of the county by the GOP continued after World War II. Dewey overwhelmed Truman in 1948 and Eisenhower defeated Stevenson in landslides in 1952 and 1956. Most agreed that these elections were best explained in ethno-religious terms or as being in keeping with traditional Schuylkill County electoral determinants. However, the presidential election of 1960 intensified these prejudices. The fact that a Catholic, John Fitzgerald Kennedy, headed the Democratic ticket in that year served to further polarize the county's Protestant-Catholic communities. The Kennedy candidacy also proved powerful enough to push the county's east European voters into an alliance with their Irish co-religionists, which had not previously occurred in the county's history.

The county results in the 1960 election definitely rotated on the religious axis. The presence of a Roman Catholic at the top of the Democratic ticket served to polarize the religious groups more noticeably than ever. The fact that Kennedy personally campaigned in the county in late October added still more fuel to the ethno-religious political fires.[22] Hence, the 1960 election intensified

TABLE IV[23]

DEMOCRATIC PERCENTAGES IN HEAVILY IRISH PRECINCTS

	1948	1952	1956	(Gov.) 1958	1960	(Gov.) 1962
Minersville 4th Ward	46.2	52.6	50.1	57.5	69.8	61.0
Mahanoy City 2nd Ward	50.2	57.4	53.9	61.8	65.6	58.8
Lost Creek #1	67.1	67.1	57.5	69.1	83.9	78.3
#2	53.5	71.5	72.0	85.5	70.8	72.5
#3	76.4	76.9	74.1	72.2	94.5	85.9
Cass Township (North)	95.6	91.4	91.0	94.8	96.6	93.3
Girardville (West)	46.6	52.4	53.8	71.6	71.8	66.5
Democratic Average	62.1	67.0	64.6	72.2	78.9	73.7

the religious base of Schuylkill County politics; Kennedy ran better in precincts where Democrats had traditionally fared well, but ran behind past Democratic totals in heavily Republican areas. This can be most clearly seen by contrasting Kennedy's percentages in Irish and German precincts with those of other Democrats who ran in the post-war era.

Table IV shows that Kennedy took Irish precincts that were in the main heavily Democratic and made them even more strongly Democratic. North Cass (the small village of Heckscherville) gave every Democratic candidate from 1948 to 1956 at least ninety-one per cent of the vote, but it afforded Kennedy an amazing 96.6 per cent (a 402 to 14 margin!). Of the seven Irish precincts listed, JFK had the highest Democratic percentage achieved since World War II in six. Computerized analyses show that JFK's "mean" in these districts was 78.9 per cent, as compared to Truman's 62.1 per cent. The only major Democrats who ran close to JFK's mean were David Lawrence (72.2 per cent) and Richardson Dilworth (73.7 per cent), the gubernatorial candidates in 1958 and 1962. Coefficient correlations comparing the JFK vote to that of the other Democrats indicate that he intensified the tendency of the county's Irish to vote Democratic.

One other point relative to the "Irish" vote is worth noting. JFK's strongest showings were in the Irish districts located outside of boroughs or in areas once known as the "patches." For example, he made incredible showings in North

Cass Township and the Lost Creek areas of West Mahanoy Township. He ran slightly weaker in the Irish precincts located in the boroughs (Minersville 4th Ward and Mahanoy City 2nd Ward). This is undoubtedly a reflection of the fact that the "township Irish" voters were more homogeneously grouped than the "borough Irish." Very few non-Irish lived within these non-urban Irish enclaves.

If JFK then managed to win bigger where Democrats usually won, he was equally adept at losing more heavily where Democrats usually lost. Table V will clearly indicate that the German Protestants rejected JFK almost as intensely as the Irish Catholics embraced him.

Kennedy ran poorest of all post-war Democratic candidates in each of the six German Protestant precincts under study. He was beaten by more than three to one margins in all of the six districts. JFK's mean vote in these German areas was a mere 18.6 per cent. The poorest showing by any of the other major Democratic candidates of the period in these districts was Stevenson's mean of 24.9 per cent in 1956. The strongest Democratic showing was Lawrence's 35.1 per cent mean in the 1958 governor's race, and with Kennedy removed from the top of the ticket, the Democratic mean in 1962 in these German precincts was over twenty-seven per cent of the vote. JFK's strongest showing was a flat twenty-five per cent of the vote. Simply put, JFK served as a negative reference in the German areas. He took areas that were Republican and made them more Republican! The rejection of Kennedy by the county's German Protestants can be further attested by the fact that the fifteen strongest anti-Kennedy precincts in the county were heavily German, and not one Catholic church was located in or near any of them.

There is little doubt then that Kennedy further polarized the voting proclivi-

TABLE V

DEMOCRATIC PERCENTAGES IN HEAVILY GERMAN PROTESTANT DISTRICTS

	1948	1952	1956	(Gov.) 1958	1960	(Gov.) 1962
Eldred	12.5	13.6	17.5	22.8	8.8	17.1
Hegins (East)	32.2	29.6	29.1	46.6	25.0	33.5
Hegins (West)	21.9	21.2	21.3	34.0	15.9	28.6
Hubley	19.5	17.6	20.1	33.9	18.9	27.0
Upper Mahantonga (Helper)	36.1	27.4	30.9	34.9	18.9	32.0
West Brunswick	32.1	27.6	30.8	38.7	24.6	25.7
Democratic Average	25.7	25.1	24.9	35.1	18.6	27.3

ties of the county's Germans and Irish. Therefore, it was crucial to his fate in the county that he run well in the east European areas (heavily Polish and Lithuanian). The east Europeans were numerous enough to have a significant effect on the outcome of the election in the county. About 7.1 per cent of the people of the county were of Lithuanian ancestry, while 5.6 per cent were of Polish origin. In all, over twenty-two thousand countians in 1960 could trace their lineage to these two east European nations.[25] Politically, the east Europeans had leaned toward the Democrats, but a sizable minority had consistently voted Republican. JFK managed to bring the east Europeans strongly into his camp, which enabled him ultimately to carry Schuylkill County by a scant 243 votes. Table VI shows that JFK far outpolled all other Democrats of his era in the sample east European districts. Hence, his appeal was to all Catholics in the county, not merely Irish Catholics.

The movement of the east Europeans to Kennedy was dramatic. In all six cases studied, he ran ahead of every other Democrat of the period. He managed to garner over seventy per cent of the votes in Girardville's East Ward—a ward which in half of the elections under consideration favored the Republi-

TABLE VI

DEMOCRATIC PERCENTAGES IN EAST EUROPEAN DISTRICTS,

	1948	1952	1956	(Gov.) 1958	1960	(Gov.) 1962
Minersville 1st Ward 2nd Precinct	52.9	59.8	61.1	69.9	81.6	62.8
Mahanoy City 1st Ward 2nd Precinct	64.2	63.5	56.8	61.9	71.3	55.0
Shenandoah 3rd Ward 2nd Precinct	41.4	40.8	43.9	53.9	71.5	52.8
Girardville East Ward	46.0	49.6	46.0	55.4	70.5	53.5
Cass Township South	72.7	74.1	72.0	77.9	80.6	66.0
Norweigian Township Currans	53.9	65.1	54.0	67.5	70.5	61.0
Democratic Average	55.1	57.8	55.1	64.4	73.3	58.7

cans. The same was true of the Second Precinct of Shenandoah's Third Ward. The Republican supremacy in the county had been predicated on the party's ability to hold a sizable minority among east Europeans. The potency of the religious issue in 1960 enabled the Democrats to temporarily cut into this traditionally Republican bloc in the Kennedy race.

The magnitude of east European Catholic support for JFK can be seen by looking closely at the returns from heavily ethnic Shenandoah in the county's northern coal belt. Shenandoah's 1960 population was 11,073; yet it had seven Roman Catholic churches which served 10,445 parishioners. These Catholics were primarily eastern Europeans. The borough had two Polish churches with a membership of over 4,500. A Lithuanian church in the town served 3,022 parishioners.[26] The returns from this predominantly east European town were staggering. Kennedy ran ahead of Nixon by 3,966 votes (4,584 to 1,617), which meant that the town voted 73.9 per cent for JFK. When one considers that JFK carried the county by 243 votes, the importance of the "Shenandoah vote" is self evident. The Shenandoah showing reflected in a magnified way the general importance of the east European vote to JFK throughout Schuylkill County.

JFK ran strongly in other boroughs which had substantial east European concentrations. Girardsville gave JFK 1,204 votes to Nixon's 486 (seventy-two per cent to twenty-eight per cent). Heavily ethnic Minersville, which had given Eisenhower a two hundred vote win over Stevenson in 1956 and generally voted Republican, voted for JFK by a 2,193 to 1,125 margin (sixty-six to thirty-four per cent). Simply put, Kennedy increased both the turnout and Democratic percentages in all the county's east European enclaves.

One obvious conclusion about the election in the county is that Kennedy's gains in the Irish areas were largely negated by the losses his faith cost him in German Protestant areas. However, the religious issue appeared to enable him to run well ahead of the normal Democratic vote in the county's east European districts. This ultimately gave him a narrow victory in a traditionally Republican county. The last Democratic presidential aspirant to carry Schuylkill County had been Franklin D. Roosevelt in his 1940 victory over Wendell Willkie. JFK's win was even more remarkable in light of the 1960 voter registrations, which numbered 27,838 more Republican than Democrats![27] Catholics who had been drawn into the Republican party by Eisenhower obviously flocked to Kennedy's standard.

The presence of a Catholic at the head of the ticket also prompted a heavier than usual turnout in a county which normally had high voter turnouts. About 85.1 per cent of the registered voters cast ballots in the county in 1960. This was over twenty per cent higher than the national turnout of 64.3 per cent.[28] Table VII indicates how much better JFK fared than did Truman and Stevenson at the hands of Schuylkill County voters.

TABLE VII

Presidential Votes in Schuylkill County 1948–1960

	1948	1962	1956	1960
Total Democratic Vote	28,202	34,688	31,483	44,430
Total Republican Vote	44,173	50,931	51,295	44,187
Democratic %	38.9	40.5	38.0	50.2
Majority	R-15,971	R-16,243	R-19,812	D-243
Total Vote	72,375	85,619	82,778	88,617

The most salient point revealed here is that Kennedy turned a 19,812 Republican majority in 1956 into a 243 vote plurality in 1960. This represented a net gain of over twenty thousand votes in a county of 103,000 registered voters. It does not appear as though a "Protestant" Kennedy could have exerted such an influence. The fact that all the other Republican candidates running county-wide in 1960 won by margins of about three thousand votes indicates that Kennedy had a unique influence on the county's voters.[29]

Many analysts feel that Kennedy's Catholicism regained for the Democrats the votes of wayward Catholics who "deviated" from their normal voting pattern to support Eisenhower in 1952 and 1956.[30] In this sense, the 1960 election "reinstated" typical voting patterns. However, this thesis appears in need of some alteration in relation to Schuylkill County. The non-Democratic Catholics in the county were mainly the east Europeans. Kennedy won their overwhelming support. These people were not only "Eisenhower Catholics" but also appear to have been—at least to an extent—"Willkie and Dewey Catholics." In brief, JFK regained "wayward" co-religionists in Schuylkill County who had defected from the Democratic ranks long before the appearance of General Eisenhower at the head of the Republican ticket. The Kennedy candidacy caused enough Catholics to vote Democratic so that the 1960 presidential race in the county "deviated" from the normal voting pattern. As Table VI points out, east European Catholics returned to their "normal" voting patterns during the 1962 gubernatorial election. This indicates that the shift in allegiance was temporary and due to factors related to Kennedy, and did not involve a basic realignment.

The magnitude of Kennedy's impact upon the county's electorate can also be attested by the extreme polarization it prompted. For example, seventy-six (35.8 per cent) of the county's 212 precincts gave Kennedy either seventy-five per cent or more of the vote or twenty-five per cent less.[28] The precincts in which JFK was beaten by a margin of at least three to one were all in the German-Protestant areas, while the precincts which gave him victories of at least three to one were in the Irish or east European Catholic areas. Kennedy's vote in Schuylkill County ranged from the 96.6 per cent he garnered in Irish Catholic North Cass to the paltry 3.8 per cent afforded him in German Protestant Eldred Township's East Precinct. Kennedy clearly had a love-hate relationship with the voters of the county.

Another way of ascertaining the impact of religion on the 1960 race in the county is to divide the county into geographic districts and determine the correlation between the Democratic voting percentage and the approximate percentage of Catholics living in each district. In Table VIII, the county is divided into twenty-one areas (the larger boroughs) which seem to best reflect a balance between political jurisdictions and definable service areas of the county's Roman Catholic churches.

TABLE VIII

Percentage of Democrats and Catholics in Selected Population Centers

District	Democrats 1960	Catholics 1960–1964
Ashland	45.4	46.0
Coaldale	59.4	51.0
Frackville	48.4	48.0
Girardville	71.2	84.0
Gordon	23.2	21.0
McAdoo	73.1	83.0
Mahanoy City	60.9	76.0
Minersville	66.0	80.0
Mount Carbon	77.6	78.0
New Philadelphia	86.8	88.0
Orwigsburg	31.7	12.0
Pine Grove	29.4	10.0
Port Carbon	48.7	31.0
Pottsville	49.5	41.0
Ringtown	23.5	20.0
Saint Clair	64.4	75.0
Schuylkill Haven	22.3	20.0
Shenandoah	73.9	77.0
Tamaqua	38.6	30.0
Tower City	23.4	22.0
Tremont	26.0	28.0

*Based on the reported membership of Roman Catholic Churches within the districts.

The extremely high correlation coefficient of over .93 strongly indicates that religion and politics intertwined quite closely in the county during the Kennedy-Nixon contest.

An objection could be raised to this ethno-religious interpretation of the county's voting behavior. The most formidable alternative explanation for Kennedy's atypical victory might be the county's depressed economic status. Perhaps the vote "for Kennedy" was in actuality a vote "against" the Eisenhower economy. Yet, the county's economic problems actually became most acute in the 1953–1955 period when the major anthracite collieries began to close. If the state of the economy was to cost the Republican candidate support in the area, it would have reasonably occurred in the 1956 election. Eisenhower, however, ran more strongly in the county in that year than he had in 1952. A rejection of Republicanism in the county based on economic factors would have also precipitated losses by the rest of the GOP ticket in 1960; this simply did not occur.

Statistical evidence also indicates that Kennedy's appeal to the voters of Schuylkill County had a religious rather than an economic base. A "partial correlation" between Catholic percentages and Democratic percentages (used in Table VIII), when "controlled for wealth," works out to an extremely high .932.[29] Simply put, this clearly shows that there was a strong correlation between Catholics and the Kennedy vote when the influence of wealth is statistically removed.

In summary, a "Protestant" Kennedy would not have carried the county in 1960. He probably would have fared no better than the rest of the 1960 Democratic ticket in the Schuylkill area. Kennedy at the head of the ticket seemed to bring an unusually high percentage of Catholics into his camp. This, plus the high voter turnout, enabled him to win narrowly. His popular majority was held down by the tradition of Republican voting in the county and by the losses his religion caused in the Protestant districts of the county.[32] The factor of religion had altered the county's voting patterns enough to allow a Democrat to achieve a narrow victory. The grave economic decline in the county during the 1950s had not proven powerful enough to enable Stevenson to overcome the area's traditional Republicanism.

The period from 1964 to 1972 saw the county's electoral trends return to a strong resemblance of the pre-1960 patterns. The 1964 Goldwater campaign was the only notable Republican disaster in the period. This, of course, followed the national results in which Lyndon Johnson attained one of the largest popular percentages in the history of Presidential elections. The county Republicans dominated the courthouse row offices, the county commissioners' chairs, and local State legislative seats. By 1970, the county's registration stood at 60,469 Republicans (sixty-seven per cent) to 29,697 Democrats (thirty-three per cent). A study of the districts used in the earlier tables indicate that the ethno-religious support given each party had shifted somewhat, but still resembled the patterns established earlier in the century.[34]

The minority Democrats still strongly controlled the Irish districts. Table IX reveals that the local Irish did give noticeably weaker support to McGovern in 1972 than they did to other candidates. Yet, they still afforded him an average of nearly sixty per cent. Even McGovern lost only one of these heavily Irish enclaves. At the non-presidential level, these districts voted heavily Democratic in every race, including the 1972 contests.

Nothing had yet occurred to move the county's Irish Catholics away from staunch support for the Democrats. Throughout this entire period, the county Democratic chairmen were all Irish Catholics, obviously reflecting recognition of the "Irish base" of the local Democrats.

The German Protestants showed even less of a tendency to "put aside" ethno-religious politics. Table X indicates that the Democrats could not even carry the sample German districts in the Johnson landslide of 1964. The county's continued domination by the GOP obviously was predicated upon this overwhelming German support.

Since the Irish and German voters showed no propensity to change their basic voting patterns, any shift away from the county's long-held electoral habits would have to be initiated by the east Europeans. Table XI shows that the east Europeans were tending to vote more heavily Democratic than they did in the 1948–1962 period. Lyndon Johnson averaged 77.3 per cent in the

TABLE IX

DEMOCRATIC PERCENTAGES IN HEAVILY IRISH PRECINCTS

	1964	1968	1972
Minersville 4th Ward	77.7	64.1	46.8
Mahanoy City 2nd Ward	72.9	62.1	54.6
Lost Creek #1	73.2	64.8	
#2	79.0	66.3	62.5
#3	75.6	71.8	(District Merged)
Cass Township (North)	96.4	94.8	70.9
Girardville (West)	79.4	69.1	56.6
Democratic Average	79.2	70.4	58.3

TABLE X

Democratic Percentages in Heavily German Protestant Districts

	1964	1968	1972
Eldred	25.7	13.5	14.9
Hegins (East)	58.8	34.7	27.7
Hegins (West)	49.5	26.2	19.0
Hubley	46.9	23.9	24.4
Upper Mahantongo (Helper)	44.6	28.1	25.6
West Brunswick	57.2	34.1	26.8
Democratic Average	47.1	26.7	23.0

sample east European districts in 1964. This eclipsed JFK's then astonishing 73.3 per cent average of 1960. Hubert Humphrey garnered a 67.8 per cent average in 1968. Even the ill-fated McGovern campaign of 1972 netted a 50.3 per cent average in these precincts. These figures indicate that the east Europeans were responding to modern economic and political forces more than their Irish and German contemporaries. Perhaps they were beginning to realize that the county was a "textbook" Democratic area.

This could also be a reflection of the fact that virtually all of the east European voters by the 1970s were at least second- and third-generation Americans and had been educated in English. They also represented the first east Europeans who were in no way "controlled" by coal interests, since no mining employment of any significance remained. They were in a sense liberated from the traditions of the early twentieth century. The east Europeans were beginning to act more similarly to voters in other impoverished, aged, social security-oriented areas of the United States.

This slight shift of east European voters away from long-held voting patterns does not materially change the fact that ethno-religious politics had reached a "fruition" in the post World War II era. Perhaps these first inklings of change are harbingers of more significant movement in the near future. Yet, it must be noted that the Irish and Germans of the county go back many more generations than the east Europeans and they exhibit much more resistance to political change. The final conclusion must be that ethno-religious politics in the county are weakening to at least a small degree, but continue to dominate the area. The last question this paper will attempt to answer is "why" this ethno-religious political base remains.

One plausible explanation is that of "internal colonialism."[36] For many years, Schuylkill County was dominated by the Reading Coal and Iron Com-

TABLE XI[35]

DEMOCRATIC PERCENTAGES IN EAST EUROPEAN DISTRICTS

	1964	1968	1972
Minersville 1st Ward 2nd Precinct	85.3	77.5	51.8
Mahanoy City 1st Ward 2nd Precinct	77.3	63.4	54.5
Shenandoah 3rd Ward 2nd Precinct	71.0	66.9	45.5
Girardville East Ward	70.9	60.4	50.4
Cass Township South	86.3	77.6	61.0
Norweigian Township Currans	73.1	61.5	39.2
Democratic Average	77.3	67.8	50.3

pany. Economic development in the county was controlled by this outside force, which monopolized employment, controlled the political elite, and influenced the development of a cultural division of labor wherein certain ethnic groups were given preference and others were relegated to the hardest, most menial and dangerous jobs. Hechter (1975) has argued that it is under conditions such as these that ethnicity becomes a major basis for political organization, a reaction against the domination of outside forces and against the resultant cultural division of labor.

Once developed, ethno-religious bases for political organization are hard to eradicate, even after the conditions which spawned them have changed. Today, anthracite production is minimal, and domination of the county by one company no longer exists. Light industry has replaced mining as the major source of employment, but the area has not recovered economically from the decline of king coal. Perhaps it is partly because the county is economically depressed and its population relatively aged that ethno-religious voting patterns are so prevalent and stable. Urbanization pressures are minimal and, therefore, so are pressures to break up and homogenize old ethnic and religious, social and political ties. Those ties may be politically stable or even passive or dormant at most times, with inertia and tradition determining much voting, the ethno-religious factor being of historical and indirect, but not direct, importance. But as this paper has shown, the religious factor at least is strong enough to divide the county along religious lines when a specific issue is made of it.

The "internal colonialism" theory used to explain the county's continued ethno-religious voting patterns is obviously strengthened by the "age factor" in Schuylkill County by the 1970s. The sharp decline in population left precincts that were much smaller in number, but basically the same ethnically. The county simply never received (in the last fifty years) a significant influx of "new" people. The decline of the county economically assured the absence of young people in the "educated upper classes." People of this type are more able to alter matters politically than the poorer, less educated older types. All of these factors combined with a notable "out migration" of the county's native young citizens to make Schuylkill one of the State's "oldest" counties by 1970.[37]

This situation produced "static precincts." In this situation, people who vote are generally the same voters who had cast ballots ten or twenty years earlier. Since the only notable change tends to be fewer ballots cast, the political tendencies are likely to stagnate in terms of Democratic/Republican percentages. The German and Irish districts of the county seem to uniquely qualify as "static precincts." Indeed, the voting patterns established in these areas generations ago appear to prevail even yet.

The fact that the county is typified by "static precincts" in which "long-term" residents cast the great majority of the ballots can be proven in another way. The county has long been legendary for its extremely high voter turnouts.[38] However, this is not typical of an old, impoverished, not highly educated region. The one "text book" answer to the county's high turnouts would be that "long-term" residents vote far more frequently than do "newcomers" to a given region. Hence, "long-term" voters residing in relatively "static" precincts decide the fate of would-be officeholders in the county. It would, therefore, be disastrous for any candidate of even the present day to dismiss ethno-religious voting determinants in Schuylkill County as things of the past.[39]

NOTES

[1]Seymour Lipset and Stein Rokkan, *Party Systems and Voter Alignments* (New York, 1967).

[2]Michael Reich, "The Economics of Racism," in David M. Gordon (ed.), *Problems of Political Economy: An Urban Perspective* (Lexington, 1977), 183–188.

[3]Nathan Glazer and Patrick Moynihan, *Beyond the Melting Pot* (Cambridge, 1963).

[4]Andrew Greeley, "Political Participation among American Ethnic Groups in the United States: A Preliminary Reconnaissance," *American Journal of Sociology,* 80:

170–204; Raymond E. Wolfinger, "The Development of Persistence of Ethnic Voting, *American Political Science Review*, 69:896–908; Michael Parenti, "Ethnic Politics and the Persistence of Ethnic Identification," *American Political Science Review*, 71: 619–635.

[5] David Knoke and Richard Felson, "Ethnic Stratification and Political Cleavage in the United States, 1952–1968," *American Journal of Sociology*, 80: 630–642.

[6] For a discussion of the continuing role of ethnocultural politics see Richard Jensen, "The Last Party System: Decay of Consensus, 1932–1980," in Paul Kleppner (ed.), *The Evolution of the American Electoral Systems* (Westport, 1981), 230–31. For a review of ethnic political factors into the 1970s, see Everett Ladd, "American Ethnic and Religious Groups," *Public Opinion*, (Winter 1978) and Mark Schneider, "Migration, Ethnicity, and Politics," *Journal of Politics*, 38: 937–962.

[7] Nineteenth-century Schuylkill County politics were analyzed in William Gudelunas and William Shade, *Before the Molly Maguires: The Emergence of the Ethno-Religious Factor in the Politics of the Lower Anthracite Region* (New York, 1976); William Gudelunas, "Nativism and the Decline of Schuylkill County Whiggery: Anti-Slavery or Anti-Catholicism," *Pennsylvania History*, 45: 225–256. For an overall view of ethnic factors in the county's electoral patterns throughout its history, see Richard L. Kolbe, "Culture, Political Parties and Voting Behavior: Schuylkill County," *Polity*, 8: 241–268.

[8] William Gudelunas and Stephen Couch, "Would a Polish or Protestant Kennedy Have Won? A Local Test of Ethnicity and Religion in the Presidential Election of 1960." *Ethnic Groups*, 3: 1–21.

[9] Gudelunas and Shade, *Before the Molly Maguires*, Chapter II.

[10] Kolbe, "Culture, Political Parties, and Voting," Gudelunas and Couch, *op. cit.*

[11] *Department of Mines and Mineral Industries*, 1961:13.

[12] *Ibid.*

[13] *Ibid.*, 101.

[14] *Bureau of Employment Security*, 1961:14.

[15] *Department of Commerce*, 1969:5. The 1960 census revealed that 38.4% of the county's labor force was engaged in manufacturing. By 1960, 15.1% of the county's workers had to labor outside the county. The census also showed 26.4% of the countians had jobs which paid less then $3,000. The median school years completed by the average Schuylkill Countian was 9.0. Only nine of the state's sixty-seven counties ranked lower.

[16] The "Spiritual Reports" of pastors of the Roman Catholic Churches revealed that in 1960 there were 79,988 "Catholic souls" in the county or about 46% of the county's population. In 1970, the report counted 67,447 "souls" or about 43% of the population. (It must be noted that these figures are conservative since people breaking away from "dues-paying" status could not be counted. Hence, the figures could be upped perhaps two or three per cent.)

[17] For details, see D. J. Kennedy, *The Official Catholic Directory*, 1961 Edition, 294–298.

[18] *Pennsylvania Statistical Abstract*, 1964–1965, 19–21.

[19] In 1961, D. J. Kennedy listed 40 Lutheran Churches in the county which still served 25,000 parishioners. A study by Douglas Johnson, Paul Picard, and Bernard Quinn, *Churches and Church Membership in the United States*, (Washington, D. C., 1974) revealed the following about the county's religious groups in 1971:

No. of Churches	Group	% of County's Population	% of Total "adherents"
80	Catholic	46.4	56.6
49	Lutheran	15.0	18.3
51	UCC	9.9	12.2
49	United Methodist	5.4	6.6
16	Evangelical Congregational	2.2	2.2
4	Episcopal	0.8	0.9
6	United Presbyterian	0.8	0.9

[20] *Pennsylvania Statistical Abstract,* 1975, 17.

[21] Information provided by the Pottsville Social Security Office. The office reported nearly 40,000 people depended on Social Security and/or Black Lung by 1980. (The age factor in the county is attributable to the fact that many young countians are forced to leave the area to find employment. The county lacks the influx of young job seekers which typify many areas of the United States. An old adage in the region can be summarized thusly—There are two types of people in Schuylkill County: (1) those who are born here, live here, and die here; and (2) those who are born here and leave here).

[22] JFK spoke in Pottsville (the county seat) on 28 October 1960. He was the first presidential candidate to ever deliver a major speech in the county. (Teddy Roosevelt had briefly appeared in the county in 1912.) The crowd JFK drew approximated 15,000.

[23] The precincts were determined as Irish, German, east European in several ways. One method was to discuss the precincts with leaders active in the 1960 campaign. The "Spiritual Reports" also were helpful in placing Catholic groups. Most of these precincts were obviously "ethnic" in other ways. For example, Minersville, 1st Ward—2nd Precinct, had two churches in it, one "Lithuanian" and one "Polish." Hence, the Catholic precincts are much more than educated guesses. The German districts were totally devoid of Catholic Churches. The fact that the votes are analyzed on a precinct rather than a borough or general township level lessens the room for error.

[24] Lawrence also was an Irish Catholic. Stevenson's means were 67.0% (1952) and 64.6% (1956). Stevenson's divorce could not have aided him politically in a county 50% Catholic.

[25] See Table II for a detailed analysis of the county's Catholic east European and other non-Irish Catholic groups.

[26] Statistics are from the "Spiritual Reports."

[27] The actual registration in the county in 1960 was 37,582 Democrats (36.1%), and 65,240 Republicans (62.7%); only 1.1% of the county's voters were registered Independent. See the *Pennsylvania Manual,* 1961:509.

[28] High turnouts typified the county in this era. For example, 77.4% voted in the low turnout 1948 election. Other turnouts were: 78.8% (1952), 84.5% (1956), 76.1% (1964), 81.2% (1968), 74.2% (1972).

[29] Republicans swept to victory in the county in races for Judge of the Superior Court, Auditor General, State Treasurer, and State Senator.

[30] Converse *et al,* 1961.

[31] Forty-one of the 212 precincts gave JFK less than 25% of the vote while 35 of 212 afforded him an excess of 75% of the vote.

[32] The formula used to statistically remove the effect of wealth was:

$$r\ 12.3 = \frac{r\ 12 - r\ 13\ r\ 23}{\sqrt{(1 - r^2/_{13})(1 - R^2/_{23})}}$$

[33] Source, *Pennsylvania Manual*, 1968. The county traditionally had a small number of "Independents" and minority party registrants.

[34] In analyzing the following three tables, it must be kept in mind that the 1964 and 1972 presidential races were landslides of such magnitudes nationally that localized studies of virtually any area in the nation would quite naturally deviate from normal trends.

[35] Few of these "Irish, German, or east European Districts" significantly changed in an ethnic sense between 1948 and 1972. However, "Norwegian, Currans" was the scene of one of the county's few suburban developments in the late 1980s. Only this district's 1972 figures were in any way influenced by an influx of people large enough to cause ethnic homogeneity to erode.

[36] For the best examples of the theory of "internal colonialism," see: Michael Hechter, *Internal Colonialism: The Celtic Fringe in British National Development, 1536–1966* (London, 1975); Ernest Gellner, *Thought and Change* (Chicago, 1969); Paul J. Nyden, "An Internal Colony, Labor Conflict and Capitalism in Appalachian Coal," *The Insurgent Sociologist*, 8:33–43; Charles C. Ragin, "Ethnic Political Mobilization: The Welsh Case," *American Sociological Review*, 4:619–655. For a study of internal colonialism in Schuylkill County, see Stephen R. Couch, "Social Control in an Internal Colony: Pennsylvania's Coal and Iron Police," presented at American Sociological Association Annual Meeting, Toronto, 24–28 August, 1981.

[37] The "old-age" factor can be best demonstrated by these facts. The county's median age by the late 1970's was nearly 38 years. By 1980, 13.1% of the county was 65 years or older. Many political subdivisions had better than 16% of their people over 65. For example, Coaldale (17.6%), Foster Township (16.9%), North Manheim Township (17.4%), Mahanoy City (18.3%), Shenandoah (18.5%). The rapid population decline until the 1970s is best demonstrated by noting that the county lost 7.5% of its population between 1960 and 1970 (160,089 from 173,027). Shenandoah alone lost 25.2% of its population between 1960 and 1970. Sources were *Schuylkill County Department of Aging* and figures from the *Economic Development Council of Northeastern Pennsylvania* (EDCNEP), "1980 Populations by Age Groups."

[38] See footnote 28 for details.

[39] A final note on the present Republican (late 1970s-early 1980s) domination of the county. It might seem that as east Europeans move ever so slowly to the Democrats, the Republican grip will weaken. However, the county still has a significant number of non-German Protestants, mainly English and Welsh. They also give the GOP strong support. This would mean that the Democrats would have to obtain much heavier east European support to end the Republican hegemony. These non-German Protestants were not analyzed in this paper mainly because they are much more difficult to isolate into nearly homogeneous precincts.

Chapter Fourteen

COMMENTARY: THE ETHNO-RELIGIOUS FACTOR REACHES FRUITION

Matthew S. Magda

Professor William Gudelunas' paper reminds one of a story which is sometimes used in discussions about the 1960 Presidential election. The story is the following: A man is asked on election eve, "Are you going to vote for Kennedy because he is a Catholic?" "No," the man replied, "because I am."[1] This anecdote and Professor Gudelunas' paper portray an important aspect of the 1960 election: Ethnicity and, in this case especially, religion had a powerful impact on how people voted. As Theodore White wrote in *The Making of the President, 1960,* "there is no doubt that millions of Americans, Protestant and Catholic, voted in 1960 primordially out of instinct, kinship and past."[2]

In his finely researched and carefully argued paper, Professor Gudelunas has clearly demonstrated that ethno-religious factors continued to play an important role in Schuylkill County politics up through the 1970s. He has further shown that there has been a remarkable continuity to county voting patterns, as the ethno-religious voting blocs established in the early twentieth century (German Protestant Republicans, Irish Catholic Democrats, and an almost evenly divided eastern European Catholic bloc) persisted into the post-World War II era. He has noted how these ethno-religious alignments helped preserve the countywide dominance of the Republican party, despite the fact that the county had social and economic characteristics which usually make this type of political entity as a "classic liberal Democratic area."[3] Finally, Professor Gudelunas has perceptively observed that since the 1960s the eastern European voters have gradually been voting Democratic more often, and that this behavior may be a harbinger of a new political alignment. This could also mean that economic issues and government social programs are exercising more influence and, consequently, that ethno-religious politics, although still dominant, may be weakening, at least to a small degree, in the county.

While Professor Gudelunas has done commendable work in delineating the ethno-religious patterns of voter behavior in Schuylkill County and the persistence of these patterns, there are areas left unexplained and others left unexplored. For example, Gudelunas noted that John F. Kennedy was able to capture for the Democratic party the votes of non-Democratic, eastern European Catholics. Yet, while we know that the religious or Catholic issue seemed to galvanize these people to vote for Kennedy, we do not know why a mutually shared religion was sufficient motivation for these people to support the Democratic candidate and draw them away from their traditional loyalty to the Republican party. After all, as Gudelunas observed, these people were not simply "Eisenhower Catholics," they were "Wilkie and Dewey Catholics."[4] There is no explanation why these people were Republican before the Kennedy election, and why it took the Catholic issue to make them Democrats. Certainly these people had voted for Protestant candidates before the 1960 election. Was there something about the Nixon candidacy, other than his religion, that may have been particularly disturbing to these voters? What was it about the Republican ideology that appealed to these eastern European Catholics before the 1960 election? We know that in other states and in other parts of Pennsylvania, these people had turned Democratic during the New Deal and had, for the most part stayed Democrats. In fact, eastern European Catholics in general are usually associated with the core of the New Deal Liberal coalition that characterized the Democratic party for decades after the 1930s.[5] Why had these people in Schuylkill County defected from the Democratic party so early? To put it another way, why did the New Deal fail to permanently overturn Republican political dominance in Schuylkill County? Was the Republican party leadership superior? Were the county Democrats inept?

Some possible answers to these questions can be devised from the research of another scholar. In a 1975 article in *Polity* entitled, "Culture, Political Parties and Voting Behavior in Schuylkill County," Richard Kolbe claimed that there was a congruence between Republican ideology and the dominant local value structure, that the individualistic culture and patriotic sentiments of the general population have nicely matched the laissez-faire ideology and intense patriotic fervor of the Republican party.[6]

In the same article Kolbe further asserted that Republicans remained dominant for reasons other than those that could be defined as ethno-religious. In particular, Kolbe maintained that Republican leadership has traditionally been far superior to Democratic leadership in the county. Here Kolbe may provide an answer to our question concerning the failure of the New Deal to permanently change the political structure of the county. Kolbe claimed that the county Democratic party, beginning in the early twentieth century, accepted a "non-competitive" and "dependent" role in county politics. He further hinted that as part of this "non-competitive," "dependent" role, local Democratic leaders have in the past struck deals with the Republican leadership whereby the Democratic leaders agreed to help hold down the Democratic vote. In return, the Democrats would be allowed by the Republicans to participate in the sharing of the various major offices."[7] According to Kolbe, this cooptation is a major reason why county Democrats failed to make a serious and prolonged effort to recruit such newly arriving Catholics as the eastern Europeans during the early twentieth century. It was always easier for the Democratic party elite to hang onto their Republican-sanctioned, "guaranteed tenure" than to risk a campaign with an uncertain outcome and one which would involve a considerably large amount of money. As a consequence, the Democrats failed to broaden their base beyond the Irish Catholics and remained the minority party.[8] Perhaps this political arrangement and cooptation was an aspect of the "internal colonialism" that Gudelunas mentioned (but does not fully explain) as a possible factor in the development and the persistence of the ethno-religious voting blocs in Schuylkill County.[9]

In demonstrating ethno-religious voter behavior, Gudelunas relied primarily upon voting results during the years 1945–1972. While one can praise his careful use of these statistics to delineate the existence of ethno-religious voting blocs, these statistical correlations tell us little about the internal dynamics of politics in Schuylkill County. We learn little about the interconnections between party leadership, programs and ideals, and the specific beliefs, values, and leadership of the various ethnic communities. We have no explanation of what was the ideological content of the Democratic and Republican parties in the county. Were these parties far apart in ideology, with one party clearly liberal and the other staunchly conservative? Or were these parties much closer in ideology? If economic and social issues were not sharply drawn between the parties, did this further the reliance upon ethno-religious issues and conflicts by each party's leadership? How did the ideological character of each party influence the voting behavior of the different ethnic groups? What specific social values of each ethnic group dovetailed with what ideals in each party? To what extent were cultural differences reinforced by ethnic political activity? How have ethno-religious factors been expressed in such areas as local school boards and at the city council or borough council level?

Other studies by political historians, such as Paul Kleppner, Lee Benson, and Michael Holt[10] have given us a picture of American politics as a matter of clashing cultural styles and world views, yet Gudelunas does not provide us with a detailed picture of these conflicts in Schuylkill County. Consequently, there is little cultural substance to the ethno-religious politics as described by Gudelunas.[11]

The 1960 election served to polarize the county's voters along religious lines and obviously is representative of a situation where religious issues become dominant. Yet we do not know when ethnic identity *per se* becomes a stronger force. For example, do German voters in Schuylkill County vote more consistently and in larger numbers for German candidates? Do the eastern Europeans and Irish show greater preference for candidates from their respective ethnic backgrounds? Some studies have shown that there is a close correlation between the ethnic background of a candidate and the support received from voters sharing the same background, and that many candidates firmly believe that they have an advantage in holding the loyalty of voters who share their ethnic identity.[12] Were these tactical considerations important factors affecting the plans and candidates of the political parties in the county?

There are other cross-currents which must be investigated before one can fully and accurately assess the role of ethnicity and religion in Schuylkill County politics, 1945–1972. One does not have to be an economic determinist to know that economic issues such as unemployment and inflation have influenced election results, and that economic class and occupation have occasionally affected voter preferences. Educational levels too have had some influence on how people have viewed government and politics. Although sometimes overplayed, national programs and major events, such as wars and scandals, have at times affected American voters. In order to gain a better understanding of politics in Schuylkill County, an attempt must be made to measure the influence of these non-ethno-religious variables upon behavior.

While voters in the county may have been profoundly influenced by the religious issue in the 1960 presidential election, it is hard to believe that the same voters would not have been affected by such issues on the local level as borough taxes, the management and effectiveness of the school system, and costs for street repairs. It is also hard to believe that these voters were aloof from such emotionally charged issues as the Vietnam War, the Civil Rights Movement, and the Great Society programs, especially during the presidential and congressional elections of 1964, 1966, 1968, and 1970. Certainly some of the above issues had some meaning for the voters of Schuylkill County. Perhaps upon investigating these issues, we would find that the same ethno-religious voting blocs persisted, but we cannot be sure of that until those areas are explored. It is quite possible that on some of these issues, one's social class, occupation, and educational level may have been much more significant than one's ethno-religious background.

There is another variable of politics which is often overlooked or ignored by political historians because this variable is difficult to study and consequently is not easily measured. Nonetheless, this variable can play an important role in determining the relative significance that ethno-religious factors versus other social and economic factors can have in elections. This variable is information. Here I am not referring to educational levels. By information I mean both the amount of information available about candidates and issues for specific elections and the quality of this information. An "informed electorate" has always been touted as a vital element for the maintenance of a healthy and vibrant democracy. Yet how well have our voters been informed about issues and candidates in the past. Is it not true that the less information available and the lower the quality of this information, the more likely it is that voters will fall back upon primordial orientations when voting? In a situation where there is abundant and sound information available, is it not likely that voters will be more influenced by issues *per se* and by the actual platforms of the parties and candidates? Before we can make accurate judgments about the importance of ethno-religious factors and their persistence in post-World War II Schuylkill County politics, we need to discover what amounts and kinds of information were available to the voters of Schuylkill County during specific elections. We need to know more about the sources and the nature of this information. This will mean that newspapers, radio programs, and television news broadcasts will have to be examined for their coverage of issues and candidates.

Nothing in this commentary dismisses the fact that ethno-religious factors have influenced politics in Schuylkill County, and throughout Pennsylvania and the nation as a whole. Certainly, enough studies have been published to verify the impact of ethnicity and religion upon American politics. Professor Gudelunas has given us an outline of ethno-religious voting blocs in post-1945 Schuylkill County. Before we can truly comprehend the role and relative importance of ethno-religious factors in county politics during this period, however, we need to have more of the historical pieces in place.

NOTES

[1] As quoted in Richard Polenberg, *One Nation Divisible: Class, Race and Ethnicity in the United States Since 1938* (New York, 1980), 164.

[2] Theodore White, *The Making of the President, 1960* (New York, 1961), 388.

[3] *Supra,* 174.

[4] *Ibid.,* 16.

[5] Norman H. Nie, Barbara Currie, and Andrew M. Greeley, "Political Attitudes Among American Ethnics: A Study of Perceptual Distortion," *Ethnicity,* I (December 1974), 317.

[6] Kolbe, "Culture, Political Parties & Voting Behavior: Schuylkill County," *Polity,* 8 (Winter 1975), 247.

[7] *Ibid.,* 259.

[8] *Ibid.,* 259–260.

[9] *Supra,* 183–84.

[10] Paul Kleppner, *The Cross of Culture: A Social Analysis of Midwestern Politics, 1850–1900* (New York, 1970); Lee Benson, *The Concept of Jacksonian Democracy* (Princeton, 1961); Michael Holt, *Forging A Majority: The Formation of the Republican Party in Pittsburgh, 1848–1860* (New Haven, 1969).

[11] In his PhD. dissertation, "Before the Molly Maguires the Emergence of the Ethno-Religious Factor in the Politics of the Lower Anthracite Region" (Lehigh University, 1975), Gudelunas provides a detailed analysis of the interconnections between ethnic cultures and the ideals and styles of political parties in Schuylkill County between 1844 and 1872. Such analysis is missing in his work on the politics during the post World War II era.

[12] Walter A. Borowiec, "Perceptions of Ethnic Voters by Ethnic Politicians," *Ethnicity,* I (October 1974), 267–268.